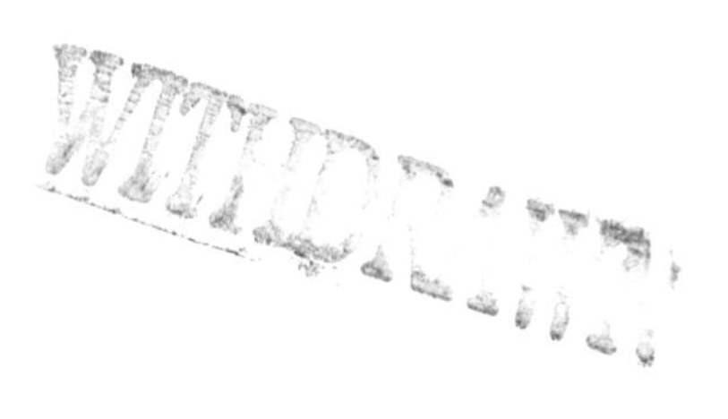

Performance Improvement

Making it Happen

Second Edition

DARRYL D. ENOS, Ph.D.

President, Achievement Associates, St. Louis

Auerbach Publications
Taylor & Francis Group
Boca Raton New York

Auerbach Publications is an imprint of the
Taylor & Francis Group, an **informa** business

Auerbach Publications
Taylor & Francis Group
6000 Broken Sound Parkway NW, Suite 300
Boca Raton, FL 33487-2742

Printed in the United States of America on acid-free paper
10 9 8 7 6 5 4 3 2

International Standard Book Number-10: 1-4200-4584-9 (Hardcover)
International Standard Book Number-13: 978-1-4200-4584-0 (Hardcover)

Library of Congress Cataloging-in-Publication Data

Enos, Darryl D.
Performance improvement--making it happen / Darryl D. Enos.
p. cm.
Includes bibliographical references and index.
ISBN 978-1-4200-4584-0 (alk. paper)
1. Organizational effectiveness. 2. Performance. 3. Teams in the workplace. I. Title.

HD58.7.E4458 2007
658.3'125--dc22 2007060360

Visit the Taylor & Francis Web site at
http://www.taylorandfrancis.com

and the Auerbach Web site at
http://www.auerbach-publications.com

CONTENTS

PREFACE

Organizations are different from each other, and yet they are alike in many ways. Organizations of various types (large versus small, not for profit versus profit driven, etc.) usually have similar challenges, problems, and opportunities that make performance improvement important to them. The specific details of their challenges, problems, and opportunities will be different between, for example, organizations doing different types of business. However, many of the basic causes of excellence or deficiency in performance are the same across different organizations with different work teams, managers, supervisors, employees, or team leads.

Just as there are similarities in the causes of differing performance levels between different organizations, there are also many similarities in what causes good performance in organizations, and what causes it in individuals or teams. This means that there are a number of basic approaches that can improve performance, regardless of what organization is involved or what it does. Further, there are basic similarities in effective approaches for improving performance for virtually any *organization, team,* or *individual.*

This is not to say that organizations and teams have no significant differences from one to the next or that these differences are insignificant in performance improvement efforts. Clearly, organizational and team performance are influenced, for example, by the personalities, attitudes, motivations, and values of their members, which vary from one organization and team to the next. They are also influenced by the types of work they do or the type of business conducted by the organization. But the perspective here is that all organizations and teams, no matter what business they are in, are in large part the creation of their team members, and improving performance comes back to their acceptance of the responsibility to do so. The characteristics and causes of effective performance in organizations and teams tend to be very similar regardless of what kind of work they do.

There are, of course, differences among organizations and teams based on what services or "products" they offer. Retailers, route businesses, manufacturers, non-profit hospitals, and universities have different products to develop and provide, and different markets to serve. However, all of them can perform better if they have a clearly defined direction — thus the importance of a strategic vision and action plan. Furthermore, the problems that occur when the strategic vision is unclear tend to be similar among organizations despite the fact that they have different products and clients.

Leadership and organizational climate will also be different from one organization or team to the next. That is why a clear understanding of the leadership style and climate is important to understanding performance. But ineffective leadership produces similar results no matter what team or organization is involved. This includes lack of motivation, poor retention of employees, and lackadaisical customer service. Thus, there are basic similarities in causes of poor levels of performance.

One similarity is that no matter what team or organization needs performance improvement, part of the approach normally should include helping people learn to work together more effectively. There are techniques for doing this that can have great impact. The variety in improvement of team performance comes from identifying which area of a team needs improvement.

A second similarity among organizations and work teams is the absolutely critical role of leadership in defining, supporting, and "institutionalizing" performance improvement. Experience shows that more failures in performance improvement efforts occur because of a lack of leadership dedication and involvement than for any other single reason. More often than not, the failure to get leadership sufficiently involved is the result of the irresponsibility of people in this profession: facilitators and consultants in performance improvement, management development, and organizational or team development. The critical success factors for improving performance in virtually any organization are discussed more fully in Chapter 3.

The book consists of three parts. Part I, Performance Improvement: Getting It Started, contains five chapters that essentially outline a basic approach to performance improvement that can be helpful to the leadership of any type of work organization (as opposed to a bowling team or a book club). Chapter 1 focuses on the scope of performance improvement efforts in the United States, with a discussion about why we are spending billions of dollars in these improvement efforts. Chapter 2 discusses performance gaps and deficiencies, indicating how leaders and managers can identify areas needing improvement. This chapter focuses on the importance of measurement, goals, key performance indicators, and models of effective performance (critical standards). Chapters 3, 4, and 5 provide greater detail in helping organizational leadership know where and what might need performance improvement. As is true throughout this book, numerous, brief case scenarios are used to help readers understand fundamental ideas regarding identification of performance deficiencies, causes, and possible solutions.

While Part I of this edition of the book focuses on "getting a handle" on areas possibly needing performance improvement, Part II is about specific types of interventions for performance improvement. The list of interventions in Part II, "Performance Improvement: Taking Action," includes strategic planning, building a learning organization, hiring and selecting people, leadership and teamwork, teams and performance improvement, and performance management systems. Chapter 12, one of the new chapters not included in the first edition of this book, focuses on project management, lean manufacturing, and Six Sigma as recent performance improvement interventions focusing on efficiency and effectiveness. Chapters 13, 14, and 15 are about developing individuals and the importance of "lifelong learning."

Part III is all about making performance improvement stable and permanent. It is tragic that so many performance improvement efforts show little or no progress for organizations. The reasons why so much time, money, and effort at improvement are largely ineffective are discussed, and we make suggestions for dramatically increasing improvement success.

There are two features to this book that should be extremely useful to readers, especially those who make decisions about performance for organizations or teams. Throughout the book, brief presentations of actual cases are used to illustrate particularly critical points. This will help those who learn best by seeing practical examples. All of the cases are ones in which our organization (Achievement Associates, Inc.) was a facilitator or a partner in performance improvement efforts. Included in this book are case studies where the results were not positive, as well as those where performance improved.

Second, each chapter includes a list of suggested action steps for leaders. If the reader is interested in considering whether performance improvement is warranted in his or her specific organization or team, or what options are available for making performance better, the suggested action steps will be useful. These suggestions are somewhat generic but will help readers think about the application of the ideas to their situations.

ACKNOWLEDGMENTS

My professional focus and motivation has always been toward the application and use of performance improvement approaches and concepts in actual situations. As a result, I have constantly sought mentors who could help me learn basic processes, concepts, and models for understanding and improving performance. Many people have been generous in providing suggestions and collaboration in my professional development and in writing this book.

Mike Weaver, President of Achievement Associates, Inc., is the most recent and perhaps the most significant mentor I have had the good fortune to work with. Mike, a technically trained engineer who decided people were a more important focus than systems, developed or helped develop many of the concepts and models contained in this book. With Mike Weaver, and with others, I have tried to give appropriate credit for their creations throughout the pages of this book.

Tony Montebello, Ph.D., was my first mentor after I changed careers to move into full-time performance development work. Tony helped me learn attitudes, behavior, tools, and techniques necessary for success in the difficult world of performance consulting. Old habits, learned through years of academic teaching and administration, died hard — but did disappear over time.

Paul Sultan, Ph.D., now retired from Southern Illinois University, was the co-author of my first published book. I learned a great deal from Paul about authoring a work that is directed toward helping those interested in performing better. I hope I have applied those lessons well in this book.

Many clients and associates have been active in discussions with me regarding both this book and its previous edition. Most notable are Jane Wulf, Rodger Riney (CEO), and Ian Patterson of Scottrade Inc. Others include Dave Vaughn, formerly of Anheuser Busch, Inc.; and Dave O'Keefe (CEO) of Landshire Inc. Others with whom I have had a valued professional relationship include Ted Perryman, formerly the managing partner of the law firm discussed in this work; Rob Anderson, formerly of Tone's Brothers; Gary Rager, of American Car Foundry; and Dr. Tom Schroeder.

Writing a book while continuing to work in the field about which you are writing has advantages and challenges. Many of the people listed above have helped me confirm or, at times, amend my perspective on what occurred in cases discussed in the following pages.

My wife Bonnie has provided personal and professional support without which this book would not exist. Our children, Tracy, Erika, and Derick, and our grandchildren Lena, Tyrese, and Alandra, along with my wife, have been supportive, tolerant, and sources of fun and relaxation. I thank them all.

Darryl D. Enos, Ph.D.

ABOUT THE AUTHOR

After completing his Ph.D. at the Claremont Colleges in Los Angeles, **Darryl D. Enos** taught at three universities, including nine years as a tenured management professor at Southern Illinois University, Edwardsville. During this period of teaching, Enos consulted with many organizations in management and related topics, including the California State Legislature, Banquet Foods, and numerous for-profit healthcare providers.

Seventeen years ago, Enos left an executive position at a St. Louis University to enter corporate consulting full time, while teaching human resources development as an adjunct professor at Webster University. During the past 17 years, he has provided organizational development support for more than 100 organizations, including Anheuser-Busch Inc., Ralston Purina, Laclede Gas, May Co., and the Department of Family Services for the State of Missouri.

During the past ten years, Enos has focused about 50 percent of his consultation on working with three companies: (1) Scottrade Inc. (a large discount broker), (2) Landshire Inc. (a food manufacturer), and (3) Alliance Credit Union. His discussion of major organizational transformation contained in Part III of this second edition is in part the result of ten years of work with each of these organizations.

Enos co-authored his first book, *The Sociology of Health Care,* with Paul Sultan of Edwardsville, Illinois. He is currently president of Achievement Associates, St. Louis, a corporation dedicated to assisting organizations of many types with efforts to improve their measurable performance.

PART I

PERFORMANCE IMPROVEMENT: GETTING IT STARTED

Chapter 1

PERFORMANCE IMPROVEMENT EFFORTS: THEY ARE EVERYWHERE

AN OVERVIEW OF TODAY'S REALITIES

This book offers concepts, models, processes, tools, and techniques for improving performance in organizations and work teams of all types. But any decision on performance improvement must be based on agreement about what performance is, and how to assess the performance of an organization, team, or individual. One of the principal concepts reiterated throughout these pages is the paramount importance of clear purpose and direction in efforts of performance management and improvement. Thus, what is it we are trying to improve? *In short, what is performance?*

PERFORMANCE DEFINITIONS

Many definitions of performance have been suggested by organizational leaders, decision makers, and students of organizational management and performance. A large percentage of these definitions focus on financial performance: profitability, gross income, or margins. One limitation to this focus is that it tends to exclude organizations, primarily government and non-profit organizations, where funding comes from someone other than the "customer," or "consumer." The principal deficiency of the financial definition of performance, however, is that it ignores the fundamental question: *What do we need to accomplish to make the desired profit, income, or margins?* To paraphrase a concept often attributed to Peter Drucker, figure out what business you are in and how to do it well, and the money will follow.

Generating income is important to success and survival, even in many non-profit organizations such as colleges and universities or cultural endeavors. But financial health as a *sole definition* of performance does not tell us what to do to be successful — something else is needed. Some organizational leaders, therefore, talk about being the best in their business. A phrase offered by thoughtful CEOs is "world class in the business of…." But the flaw with this definition of performance is the same as with exclusive focus on profit or income as a way of defining performance: *What do we need to do to become "world class?"*

Some university textbooks written with adult students in mind who are also managers get closer. One outstanding book on organizational development (OD) says that OD is oriented toward improving organizational effectiveness, and that "an effective organization is able to solve its own problems and focus its attention and resources on achieving key goals." (1) This definition is a start at defining performance, but it assumes that the goals are clearly defined in the organization. Many organizations, of all sizes and types of business, have few if any stated performance goals, and those that do often keep them very general and vague.

The recent, very popular book entitled *Good to Great* by Jim Collins uses the ratio of cumulative stock returns for corporations compared to the general stock market. (2) In addition to excluding not-for-profit organizations, this approach makes financial success a kind of barometer of "great" as distinguished from "comparison" companies. To his credit, however, Collins goes much further, and the majority of this valuable book is a discussion of the factors Collins and his team believe have made the financially successful companies "great." These factors include organizational dimensions such as leadership style, hiring the right people, and having a culture of discipline. These and other factors regarding how organizations and teams are built and managed are the focus of *performance improvement* efforts.

So what is a useful definition of performance that can help us build and manage the organizational factors that result in successful organizations? After a good deal of experience with a lot of organizations and a great deal of research, the best definition we have found is this: *"performance is the definition and progressive achievement of tangible, specific, measurable, worthwhile, and personally meaningful goals."* (3) A few key points will help clarify the meaning and benefits of this definition:

1. Many organizations, teams, and individuals, both at work and personally, do not have well-defined, specific, or measurable goals. It is common to find smaller organizations, perhaps 20 million in gross

income, with no goals other than a structured budget. Even the budget might be poorly defined. Larger organizations (Fortune 500 companies, for example) often have very vague and general goals or objectives at the corporate level, and provide little specific direction to those charged with achieving these targets. A classic phrase known to many managers throughout the United States clarifies the issue here: "If you can't measure it, you can't manage it." More about that later.

2. Setting specific and measurable goals forces us to decide what we are really trying to achieve, and how to know if we get there. Measurement in work situations, especially in areas other than finance or production, can be scary to managers and employees for reasons ranging from difficulty in defining "soft goals" to fear of failure. However, as discussed in Chapters 2 and 5, goals can be created in a way that provides clear direction and increased chances of performance success, even in areas such as management style or customer service.
3. The fact that goals should be "personally meaningful" is probably the most critical point in the above definition of performance. Goals should not just appear; nor does having goals guarantee commitment and motivation to achieve. To get commitment and motivation, organizations need to start with the question, *"Goals to do what?"* The initial answer, which can increase commitment, is: *"Goals to accomplish the vision, the drive for success, and the desire for competence in those involved in the team or organization."* This leads to the importance of organizational strategy, which is discussed in Chapter 6.

POWERFUL NEED FOR PERFORMANCE IMPROVEMENT

Efforts at performance improvement in organizations, teams, and individual employees are everywhere. Training and development, only one of the major efforts at performance improvement, is estimated to have cost U.S. companies more than $55 billion in 1995. The American Society for Training and Development reported that a "broad cross section of organizations" spent an average of $955 per employee on training in 2004. (4) A 1994 survey done by AMA/Deloitte & Touche demonstrated that 84 percent of the companies included in the study had change initiatives underway, 68 percent had formalized change management programs going, and close to 50 percent of the companies had three change initiatives in effect. (5)

Many social and economic factors led to this huge number of performance improvement initiatives, with a major increase in those initiatives in the 1990s. These social and economic factors have increased in intensity in the past decade, meaning that the efforts at performance enhancement have also increased in recent years. The list of factors creating this powerful need for performance improvement varies with the individual leader, but what influences most leaders to want improved performance is usually one of the following:

1. Increased *competition* as a reason for concern about performance is on everybody's list, especially organizational decision makers facing intense and worldwide competition. The competition factor has intensified with the onset of NAFTA and the economic surges of India and China. A few years ago, this author discussed a set of performance issues and some possible improvement efforts with a small construction company. The significant information on that company included:
   a. Owner-operated company in a small municipality within a larger city in the Midwest
   b. Products and services:
      i. Constructed steel beams as reinforcement for commercial buildings
      ii. Delivered and helped install those beams
   c. Markets:
      i. Construction companies
      ii. Commercial architects
   d. Structure:
      i. Owner operated
      ii. A total of approximately 20 employees
   e. Primary issues:
      i. Intense competition and the need for a redefinition of their marketing approach
      ii. Some problems with leadership and management style

   In discussions with this group, it became clear that their primary concern focused on their new, intense competition. Their competition came mostly from Pacific Rim companies, providing products in the middle section of the United States. The inexpensive labor permitted the foreign competitor to maintain low steel beam costs even after paying for shipment from the Pacific Rim. The "one world economy," which has so dramatically impacted the United States through competition in the American auto industry, is with us to stay — from manufacturing, to services, to education, to high tech. Performance improvement is a requirement for survival in today's world of intense and growing competition.

2. An increase in *customer knowledge and demands* also drives the felt need for performance improvement among many organizational leaders. This increase in customer knowledge and demands is fed by easier access to information in today's networked world and by the intensity of competition, which gives customers so many options. In addition, the recent popularity of quality programs taught at least some customers, and sometimes their suppliers, that quality means doing what is right for the customer at all costs.
3. *Rapid technology changes,* and a rapid and increasing pace of change generally, often lead organizational leaders to feel the need to make efforts aimed at performance improvement. This need takes many forms: "our customers expect us to have and use this technology"; "the competition is getting it and will be ahead of us"; or "this technology will make our team more effective when we learn how to make the best use of it." Improved use of technology might mean faster responses for a call center or better technology use in a manager training program.

   In some situations, achieving the most recent technology becomes an end in itself, rather than a significant component in a broader performance improvement initiative. Thus, a company such as Scottrade Inc., an online equities trading company, makes technology a constant priority because it is how more than 95 percent of its customers buy their services. In cases like this, maintaining and improving technology is central to their business performance. But whatever its uses and importance, major technology changes often means training or restructuring to increase or maintain performance.
4. *Human resource needs and desires,* including people from all levels and units of the organization as well as important stakeholders, are often factors in motivation for performance improvement. This includes a number of well-known human resources-related trends, such as: leaderships' desire to build shareholder value; the impact of downsizing and mergers; the efficiency drive to do more with less; changing attitudes toward work (generation Xers and Millenials); the emphasis on the "knowledge worker"; and the shortage of a skilled labor supply, meaning that jobs often go unfilled and that others need to be more productive to cover the work.
5. Some experts on organizational behavior believe that human beings have a powerful *need to be competent.* Authors such as Jay Hall, Ph.D., see that need to be competent as the basis for our progress historically. "If excellence is our goal, we already have what it takes to get there. We are a people with both the desire and the

ability to do good work! We are competent." (6) The need to feel and be competent is a motivator for many people. Others seem to have misplaced it or had it driven out of them. Still, when people are motivated to get better at what they do, it often leads to considerations of performance improvement initiatives.

6. Incredible and growing *knowledge availability* — ranging from information about customers or markets and market niches, to competition, to new technology, to suppliers and their wares — often produces a strong drive to learn this knowledge and integrate it into performance. Because of the abundance and array of information and the constant growth in the supply of the knowledge and data, organizations often attempt to make use of the information. Managing the knowledge and data possessed by the organization requires learning opportunities. We discuss this more fully in subsequent chapters.

LIMITED SUCCESS IN PERFORMANCE IMPROVEMENT

With so many powerful social and economic factors motivating organizational leaders to look for performance improvement in our organizations, why do both experience and research support the conclusion that "most companies have achieved only modest success in change management?" (7) One of the primary purposes of this book is to identify how and why so much of our effort at performance improvement either fails or has very limited success. A second primary purpose of this book is to demonstrate how performance improvement initiatives can be made more successful.

Success or failure in performance improvement efforts *begins with the reasons why organizational decision makers decide to get involved in the first place.* The early discussions and thinking of organizational decision makers about performance often set the direction and tone of what, if anything, is to be done. Sometimes, performance improvement consultants, either from within or from outside the organization, are involved in early decisions, and therefore have the opportunity to clarify the perspective and intentions of organizational/team leaders. Other times, these consultants may be forced into the role of the "helping hand" and told what to do. If they accept the helping hand role, clarifying the motivations of organizational leaders in seeking performance improvement may be a lost opportunity. Whatever the process for early decisions about performance improvement, the motivation and commitment of the leadership should be clarified early to increase chances of success.

A number of years ago, we were involved in discussions concerning leadership development with a highly placed executive of a very large communications company.

Organization: A Fortune 500 Communications Company with Worldwide Reach

- Products and services:
  - Communications services
  - Communications equipment
- Markets:
  - Households
  - Businesses of all sizes
- Structure:
  - Regionalized structure with a good deal of local operational autonomy
- Primary issues and problems:
  - Apparent need for improving management or leadership style among a group of managers
  - Vagueness about why the decision maker was interested

In initial conversations, the motivations of the leader of the group of managers for wanting to improve performance of his team were difficult to uncover. Vague phrases pertaining to the need for improved leadership style, both from the primary decision maker and from others in his group involved in early discussions, were the primary data provided. The collaborative or "partnering" approach so valuable in performance improvement did not occur. Communication was not open. The leader of the group kept his motivations for management development a secret. (8)

During the performance improvement interventions, additional data about the management group was collected, and some basic conclusions became clear. First, the top leader himself was a major part of the leadership problem, and others knew it. Second, his motivation for performance improvement was primarily to meet a commitment he had made to his own manager about improving his team. In short, he had not really thought about what his group needed for performance improvement and had little interest in collaborating to determine those needs or to help make his team better. His motivation was exclusively to satisfy his boss.

Throughout this book, there is the stated position that the commitment and involvement of leadership is the most critical element in whether or not performance efforts, when initiated, are successful. Support will be provided for that point of view a number of times. In the context of this discussion, however, one basic point is important. If the organization or team leader is not motivated by authentic interest in and commitment to performance of his or her group and is not open minded about doing what is best for developing that group, performance improvement efforts have little chance to work. Being authentically interested in and committed to performance is essential to being open minded about what to do and committed to search for and initiate the best methods of performance improvement.

SPECIFIC REASONS WHY LEADERS START PERFORMANCE IMPROVEMENT

Anyone reading the brief description of the above case describing the Fortune 500 communications company might conclude that we were lucky to have leadership support, whatever the leader's motivations. This gets us directly to the important issue of why organizational leaders get involved in performance improvement and the impact it has on success. A small number of primary motivations for focus on performance improvement are frequently found among organizational/team decision makers:

1. *The leader said so* is a common reason why organizations start a structured program for improving their performance. Leadership support is absolutely necessary for performance improvement in any organization — but it is not sufficient by itself. The leader's support must be such that he or she is dedicated to doing everything necessary for making performance improvement happen and be retained. In the communications case cited above, the local leader's own "boss" wanted him to work on management style, and that was his primary motivation. The managers below the local leader got involved in the management development program because their local "boss" said it was required.

   There are a number of negative consequences from engaging in performance improvement solely because the leaders said so. First, organizational leaders may or may not understand what is causing performance deficiencies, but they almost never know how to improve it. When accepting a performance program is based only on the *boss said so,* the strong tendency is to not only accept the preliminary judgment that improvement is needed, but to buy into preliminary decisions about the specific nature of the problem and what is needed to fix it.

   Second, someone must confront and clarify the leader's motivations to make sure dedication to performance improvement is sufficient to sustain it. So the requirement is that someone from inside or outside the organization needs to help the leader make sure that there is sufficient dedication to making performance improvement happen and be maintained. And finally, the perceptions of the leader regarding the team or organization should obviously be considered, but cannot be taken as the final word.

As we discuss in later chapters, the diagnosis of the performance situation and what needs improvement is the most critical stage in efforts at developing a team or organization and should not depend solely on one person's point of view. When people of high position within the organization/team are motivated to make "something happen" with their group, the tendency is to accept their judgments about what to do without sufficient diagnosis of what is really going on in performance.

Confronting a top decision maker or CEO in any organization about their motivations and perceptions is risky business. It is risky whether the person doing the confronting is from inside the organization or from outside it. The key here is why and how the confrontation is conducted. Good intentions go a long way.

2. *There is money for developing our organization or team* is a second common reason why decision makers consider performance improvement. This reason for engaging in performance improvement is a lot like "the leader said so." It sounds good but has its dangers if diagnosis stops there. Developing organizations, teams, or individuals costs money, many billions of dollars each year in the United States alone. But being able to meet the practical necessity of paying for the efforts is only a start. The key step is to thoroughly diagnose the performance situation, identifying problems, issues, and causes. We discuss much more about this in Chapter 3.

   The reverse situation of "we have no money" also occurs frequently, and lack of money can end discussions about performance improvement. Because of the difficulty in measuring return on investment for performance improvement efforts, these efforts are usually seen as cost centered, and are therefore a luxury to be forgotten when money is tight.

   Organizations have budgets, and budgets are set up on a time limit, usually one year. It is tempting to fall into the "we better spend it or we will lose it" trap. The availability of money, like the support of the leader, is desirable but cannot be the sole or primary motivation for engaging in performance improvement.

3. *Others are doing it,* sometimes referred to as the "fad" motivation, is subtly powerful in decision making of organizational or team leaders and performance improvement facilitators. Sometimes, fad as a reason for doing something comes from external consultants who have the "latest and greatest thing" that they want to sell. (9) Just as often, the ideas about what "others are doing" come from managers within the organization and CEOs who talk to and influence each other in various meetings and encounters.

> CEOs and top managers are often a close-knit group within a geographic area or an industry. They spend a lot of time talking with each other at professional associations, during Board of Directors meetings, at the local bank, and in social gatherings. They are often a valuable source of ideas and information to each other, but they can also have undesirable influence. Because managers share leadership responsibilities, they have empathy with each other and experience similar issues. Only infrequently, however, do they know how to diagnose or improve performance. That is even truer if the pre-emptive diagnosis is done during "happy hour" or on the basis that "it is helping us and it may help you."
>
> There is a special version of the "fad" — the idea that a performance improvement effort worked before for the organization/team involved, and therefore should be done again. As is the case with other motivations for decision makers' involvement in performance improvement, the problem here is the tendency to jump to action, therein minimizing diagnosis. Again, the result is insufficient understanding of the performance status, deficiency, causes, and possible solutions.

Dedication and active support from organizational leaders is absolutely critical to the success of performance improvement. Having the finances to support the program is also critical, and there is nothing inherently wrong with a performance improvement program being popular. The question is whether these motivations dominate the initiation of the program, and result in inadequate diagnosis of the issues and opportunities. The specific intervention aimed at performance improvement — whether it is restructuring, strategic planning, a major training and development effort, or some other intervention — cannot succeed if it is aimed at the wrong target.

Knowing where the organization's performance currently is and identifying successes and deficiencies clearly is a necessary critical early step in performance improvement. Questions that must be answered include the following:

- How clearly defined and stated is the business strategy?
- Is the organizational structure effective?
- Are the human resources being well managed?

This is the essence of *diagnosis*.

We started this chapter by defining performance as "the definition and progressive achievement of tangible, specific, measurable, and personally meaningful goals." If the performance improvement efforts themselves lack defined, tangible, specific, measurable, and meaningful goals, then those efforts at building organizations and teams are doomed to partial success at best and failure at worst. A brief actual case study can help illustrate this principle.

Organization: Small Consumer Services Company

- Background of the organization:
  - Small consumer services company operated by founder's sons
  - Owned by two brothers who were very concerned about the company's future
- Products and services:
  - Servicing and installation of HVAC systems
  - Maintenance contracts on HVAC systems
  - Higher technology air quality systems
- Markets:
  - Households
  - Commercial buildings
- Primary issues and problems:
  - Decline in gross income over recent years
  - Issues of economic decline in their geographic service area

In the original conversations with the owners of this 40-year-old company, it became clear that they had decided on the nature and solution to their problem. They believed that they needed to improve their customer service to get back their lost business and to keep what they still had. After lengthy discussions, including a review of the way they had defined their customer groupings (markets), it became clear that the urgent need was to expand geographic markets, find ways to provide services further from the home office, and improve selling efforts. After more discussions, the open-minded owners changed their opinions and agreed to efforts aimed at these strategic and selling performance needs. Customer service was not the urgent need. Providing better service to a market shrinking because of demographic shifts would not solve the basic problems.

The owners of this small company wanted to be successful, but they were not sure how to define that or how to get there. *Increased clarification of their vision of the company* — what it should be like in the future — was essential if they were to succeed. An early step was to define the specific goals of the performance improvement efforts. They were as follows: redefine expanded markets to be served, develop a strategy for selling effectively to penetrate those markets, and find approaches for distribution of the services to be provided to the expanded markets. Once this "strategic vision" was established, other efforts at performance improvement were initiated by the company, including customer service for the newly defined markets.

A PROCESS APPROACH: THE FIRST LOOK

The main point of this case, discussed again later, is that these leaders needed to clarify their "vision" or expanded mental picture of the desired and intended future of the company they owned. However, they had not succeeded in doing it by themselves. We all tend to benefit from discussions of important topics with others who are well intended. Organizational leaders can benefit from open discussions about performance, either with people working with them or with improvement "experts." It is often very productive to use a collaborative process to define the leaders' vision and understanding of the organization's needs.

When discussion with others is going to occur, the organization or team leader should bring dedication to candor and willingness to share both hard data about the organization and impressions and tentative conclusions about current and future performance. The internal or external performance improvement "expert" should bring, or be quick at developing, models of effective performance for the areas of concern, as well as a great ability to listen and collect information about the situation. Much more about this in Chapter 3.

As the issues, needs, and opportunities are clarified, possible interventions to improve performance will emerge based on this diagnosis. Many interventions are available and are discussed throughout this and many other books on performance improvement and organizational development. However, the most critical stage in performance improvement, perhaps surprisingly, is not the intervention, but rather the *diagnosis.*

Organizational leaders facing participation in diagnosis and possible performance improvement efforts have a tactical decision to make about the scope of the performance improvement effort. A simple distinction is between the *micro level* and the *macro level.* Micro level is when the leader thinks the focus should be limited to developing the supervisory team, sales team, customer service group, or MIS department. Sometimes, micro focus includes developing the performance of a single individual who is important to the organization.

Issues of performance do at times appear particularly critical in one area of the organization, and therefore focus on the micro level can be appropriate. One challenge of micro focus, however, is that organizations are systems, and working to improve performance in one area may have unforeseen and even negative consequences in another. For example, as experience has shown a number of times, improving production or sales in a manufacturing company may cause problems unless the logistics department responsible for supplies of raw material is made aware of the changing need for these materials. Another example would be that improving supervisory knowledge and skills to increase the amount of on-the-job training provided by supervisors probably means that those hiring new supervisors will need to clarify that this is expected also of new supervisors.

In the best of all worlds, focusing on the macro level is more desirable because it offers the opportunity to move all facets of the organization ahead. This does not mean that "we have to do everything at once," as many CEOs fear. Rather, it means that an organizational performance improvement plan needs to be developed that involves coordinated and timed efforts in each area of performance improvement where the need exists. We discuss this more fully in the final chapter of this book.

SUMMARY

William James, a famous American philosopher in the early 1900s, is quoted as having said that the way we begin an endeavor is the most critical factor in its success. Effective diagnosis of the performance needs, deficiencies, and issues is essential to success in performance improvement. But effective diagnosis requires candor and good intentions on the part of organizational or team decision makers.

Organizational decision makers and performance improvement consultants alike need to pay attention to the way they begin the important endeavors aimed at performance improvement.

A belief often expressed by organizational leaders as well as consultants is that the motivations and attitudes of those deciding on performance improvement are not that critical. After all, "something is better than nothing." Added to this is the well-accepted notion that organizations, teams, and people have a natural tendency to resist change, so even the best performance improvement efforts will have limited success. (10) Thus, why make a complicated and difficult process more so by requiring involved collaboration and diagnosis?

Misdirected efforts at performance improvement are often the cause of, or at least major contributors to, resistance to change. The use of *key performance indicators,* discussed fully in Chapter 2, helps determine if the direction for improvement efforts is correct. When efforts are directed at the wrong issues, involve the wrong people, or do not have leadership support and participation, time and money are wasted, and frustration among participants can skyrocket. One sad version of this is the common attitude by members of the organization/team that the performance improvement effort is something to be endured, probably doomed to be ineffective. This is even stronger if the current performance improvement program is simply the most recent in a string of "failures." When this happens, *something is often worse than nothing.*

SUGGESTED ACTION STEPS FOR ORGANIZATIONAL OR TEAM LEADERS

Discuss the following topics with the top decision makers within your organization/team.

1. Do we have reasons for concern about performance in our organization or team?
2. Is the vision of our organization's future clear to all involved in making it happen? Is it defined in a way that is specific and meaningful to us? Or, have we become satisfied with a two-paragraph mission statement that does not clarify what we are about?
3. Do we have the expertise to clarify how well we are performing and to identify barriers to performance in reaching our goals? Do we have people on staff with expertise on performance improvement efforts? Who should we get involved in deciding what areas, if any, need improvement?

END NOTES

1. Cummings, Thomas G. and Worley, Christopher G., *Organizational Development and Change.* South-Western College Publishing, 6th edition, 1997, p. 3.
2. Collins, Jim, *Good to Great,* Harper Business, 2001.
3. This definition is used by Achievement Associates, Inc., in a number of our performance improvement efforts. As is often true of intellectual capital, many organizations and performance experts have contributed to formation of the ideas.
4. Bassi, Laurie and Van Buren, Mark E., State of the Industry Report, ASTD, 1998, p. 26.
5. AMA/Deloitte & Touche LLP, *Survey On Change Management,* 1994, p. 2.
6. Hall, Jay, Ph.D., *The Competence Connection: A Blue Print for Excellence,* Woodstead Press, 1993, p. xii.
7. AMA/Deloitte & Touche LLP, *Survey On Change Management,* 1994, p. 3.
8. Robinson, Dana Gaines and Robinson, James C., *Performance Consulting: Moving beyond Training,* Berrett–Koehler Publishers, 1996.
9. Schermerhorn, John R., Hunt, James G., and Osborn, Richard N., *Managing Organizational Behavior,* John Wiley & Sons, 1994, p. 495.
10. Robbins, Steven P., *Organizational Behavior: Concepts, Controversies, and Applications,* Prentice Hall, 1996, p. 723–ff.

Chapter 2

PERFORMANCE GAPS AND DEFICIENCIES: AN OVERVIEW FOR FACING REALITY

Chapter 1 stated that performance is the "definition and progressive achievement of tangible, specific, measurable, and personally meaningful goals," and then asked "goals to do what?" This had to be troublesome to readers concerned with moving straight through to understanding performance improvement. The reality is, however, that working to improve performance in any organization or team often involves the same back and forth, or iterative, steps. Do the leaders know what they want to accomplish; do they have a strategic vision? Sometimes yes, sometimes no. To clarify the issues here, we will put these concepts together in a way that can provide a framework and techniques for guiding discussions and decisions about current performance and needed improvements.

The first step is to accept the fact that all organizations, no matter how successful, have problems. The question is whether these problems need to be addressed and improvement accomplished. A problem is anything that delays, stops, or minimizes the achievement of an important, "worthwhile" goal. Some problems are more positive than others. An example of a positive problem is as follows: We have an opportunity with a major new client that can dramatically improve our business income. Problem? They want a written proposal in detail by the end of the week.

How about a negative problem? We got that major new business, stopped selling other new business for lack of time and resources, and now the major new client has terminated future business with us. This is an everyday issue with businesses of all sizes, and it occurs also in not-for-profit organizations.

Sometimes, performance improvement starts with a targeted program for dealing with a problem that is limiting achievement of a worthwhile goal: developing a new product or service, installing upscale technology, reducing employee turnover rate, etc. Somewhat surprisingly, as often as not the *real* issue is that "the worthwhile goals" have not been agreed to by enough people to generate any action. Part of the early analysis in performance improvement must answer this question: Do these people agree on the specific goals they are trying to achieve? If not, we need to start there. If yes, they have agreement on specific targets, the next step is to identify the problems that are working against their success.

How can any leader or manager of an organization not know what needs to be done, what the goals are? How can any leader or manager of an organization not accept the fact that their organization has some things that are problems that are getting in the way of the best success they can accomplish? The following case study of a profitable partnership from the healthcare industry helps clarify the issues here.

Organization: Chiropractic Partnership

- Products:
  - Full-service chiropractic
  - Diet and exercise services
- Markets:
  - Families
  - Athletes with injuries
- Partnership:
  - One founding partner
  - One new partner joining the practice in the past year
- Primary issues and problems:
  - Maintaining long-term focus on business development efforts

This partnership, housing six chiropractors, has been in business successfully for more than 16 years. Its success has been due largely to the caring orientation of the original partner and his ability to hire employees who also provide great patient service.

While somewhat willing to engage in performance improvement efforts, the founding partner's following statement was significant: "There can't be much wrong here; we have been successful for many years." Effective leaders come to recognize that, no matter how successful their organization is currently, maintaining successful performance requires continuous effort at improving all the factors that lead to that success. To take some license with Sachel Paige's famous quote, don't stop to look back. If you do, they will have time to be gaining on you.

LEADERSHIP DEFINITIONS OF THEIR ORGANIZATION'S PERFORMANCE: THE IDEAL CASE

1. In an ideal world, the leadership of any organization or team would have clearly defined their strategic vision with all the elements described later in Chapter 6. They would have answered questions such as the following: What business are we in? What do we do for whom? How can we be effective and efficient in achieving the vision we have laid out? In this ideal situation, the strategic vision is defined enough that strategic and daily operational goals can be determined, clarified, and communicated.
2. In this ideal situation, standards, goals, productivity measures, or other models of effective performance have been defined at the operational level. (1) These operational goals have been established as a result of the strategic goals. Everyone understands his or her role and responsibilities — who is responsible for what and when.

Unfortunately, this ideal situation usually does not exist.

At the leadership level, many organization and team leaders have only vague ideas about what they want their organizations to be, now or in the future. In fact, many leaders are so wrapped up in solving daily crises and operational issues that they have not thought through what they desire to achieve. Even if they have a "plan" or vision of their desired organization or team, they often do not communicate it clearly to others. In fact, many organizational leaders behave as if they really believe that "as long as I know what we are trying to achieve here, that is all that really matters." Therefore, leaders often live largely alone in their own vision and do not have a chance to gain support from those who could work with them to accomplish what needs to be done.

This frequent deficiency in leadership vision, either through lack of clarity and specificity or because of the unwillingness or inability to communicate it to others, is not a condition reserved for any particular type of leader or organization. This sad and ineffective condition can be found in organizations that are for profit and not-for-profit, with leaders of all races and both genders, and among those from all age groups. One of the basic functions of management is to provide direction, but many leaders only provide marching orders. Sometimes, even these marching orders are unclear. (2) Marching orders are usually unclear if only the leader knows the final destination.

This deficiency in defining and communicating the strategic vision can come from a lack of appreciation of the role of leaders in setting the long-term direction of the organization. Sometimes, even when they appreciate the strategic role of leadership, leaders tend to have the mistaken belief that performance will occur best when they keep their vision of the organization to themselves, as their sole prerogative. The belief in this case is that the workers do not need to know the direction of their organization to do their work. In fact, some leaders believe that communicating the direction of the organization or team is a distraction and confusing to workers. There is a lack of understanding on the part of some leaders about how sharing organizational direction is essential in today's rapidly changing realities. In short, there is little appreciation of the potential motivational impact of what Drucker has called "policies to make the future." (3)

Even with the best intentions, some organization and team leaders have not figured out *how* to develop and communicate their mental picture of the organization, as they want it to be. The following brief case study helps clarify the point.

Organization: Webster University

- Products and services:
  - Undergraduate and graduate-level collegiate education
  - Related services such as books, residence facilities, etc.
- Markets:
  - Graduate students, usually employed full time
  - Undergraduate students with recent high school graduation
  - A large number of foreign-born students
- Structure:
  - Main campus provides direction for distant sites
  - Numerous distance-learning facilities inside the United States and also in a number of other countries
- Primary issues and problems:
  - Executive leadership at the central location of the university decided that each major unit needed a written plan for the future
  - The primary marketing unit of the university had recently been restructured to include some distinct functions previously operating independently of each other
  - Common team direction and interaction was insufficiently established

The leader of the marketing unit was a highly intelligent, educated, and insightful person who only needed some help in clarifying the primary issues/problems and deciding how best to improve the situation. After a great deal of consideration and diagnosis, she decided to engage in a process for developing a shared plan with 20 of her administrators and staff. The objective was to develop a written plan that would be useful for building teamwork, defining and evaluating performance, and providing rationale for future budget requests.

The objective of developing a written plan was achieved. The challenge then became the standard "How do we get the strategic vision and related goals into operation?" (4) However, the performance improvement initiative had succeeded in getting the leadership vision developed and shared by the top team. That vision included specific major activities for each of the organizational units and actions necessary to further develop the marketing unit.

PERFORMANCE AT THE OPERATIONAL LEVEL

The leaders' thorough and specific vision of the organization (in the above case, a department or independent team) is the beginning basis for defining performance, measuring it, and deciding if improvement is needed. The leaders' vision should be clear enough to lead to specific goals that are themselves measurable, that therefore define performance. But that detailed and specific vision also provides the basis for performance definition *throughout* the organization or team, as the strategic vision and goals are defined and applied throughout the entire organization. Everyone needs to know specifically what the organization, company, university, or department is attempting to achieve. Then the possibility exists that everyone can work for the shared vision.

The beginning of this book discussed the need for leadership to provide a detailed, specific, and thorough vision that they use as the start of defining performance for themselves and others in the organization. Who will be involved in defining this vision will vary depending on the nature of the business, as well as the top person's leadership style, personality, and attitudes. Some leaders want few others involved; other leaders want to include a large number in deciding on future direction. The best guiding principle is this: *include those persons whose support and acceptance are essential to achieving the vision.*

Many writers and consultants expend a good deal of effort in discussing the relationship between participation in planning specific goals on the one hand, and the acceptance of the plan and goals, plus commitment and motivation to achieving the plan and goals, on the other hand. (5) These are important questions because they focus directly on whether involving others in defining the strategic vision and resulting goals is worth the effort. A few basic principles, supported by lengthy experience and substantial research, are warranted here (6):

1. People's attitudes influence or modify behavior, although there is some debate on whether attitudes cause behavior.

2. Participation in planning and goal setting increases acceptance of those goals.
3. Challenging goals increase performance, as long as the goals are seen as achievable. Participation in goal setting, when done correctly, can lead to very challenging goals, acceptance of those goals, and increased performance.
4. Some people are more motivated by participation in setting goals in which they are involved; others appear to be at least equally motivated by goals that are assigned. It is not easy to know in advance which group is which.
5. Specific goals are much more motivational than vague or general standards — for example, "improve your performance" (vague) compared to "increase production by 2 percent next week" (specific).
6. Measurement and feedback to teams and individuals regarding goals for which they are responsible modify behavior and probably increase performance over what it would be without measured feedback.
7. The above principles work about equally well with individuals or teams.
8. While job satisfaction does not automatically produce performance (i.e., some satisfied employees are *not* performing at high levels), good performance usually increases job satisfaction.
9. There is a direct, positive relationship between job satisfaction and retention of employees.
10. In summary, goals and feedback on these goals to those working to achieve them increase performance and job satisfaction. This can result in better employee retention. However, a "satisfied" employee will not necessarily be a productive one without being managed correctly. In Chapters 5 and 9 there are discussions regarding the role of goal setting in effective management.

While these are important principles, one further point is critical. A leader of a team or organization needs the support and effort of those working for him to increase the possibility of having performance excellence. It is true that some employees will feel committed and motivated when they are involved in planning (or when someone at least explains the plan to them), while others may not feel this way. Commitment from some employees is a start. Others who are less committed and less motivated can be worked with as time goes on.

The *style of involving others in defining future direction* influences acceptance and motivation, as does the receptivity of the employees. Involvement in defining the vision or the goals of the organization or team will best produce acceptance or motivation if participants see that involvement as real, not plastic. "Plastic" participation means that the participants have no real influence over significant decisions and goals impacting the work environment or success of the organization.

Once the detailed vision of the leader is defined and communicated, it is time to set the specific goals, establish standards, and identify models of effective performance for people throughout the organization or team. The process for this, and a lengthy discussion of goals, *key performance indicators*, and standards, is contained in Chapters 4 and 5. Here we focus only on a brief definition and a few examples to clarify these important concepts.

1. A *goal* is an end result. There are many different types of goals, for example, profit margin, the number of people to hire, or acceptable return rate on a product line sent to customers.
2. A *standard* is a very specific requirement, often in areas of production such as quality measures for a tangible item, or in areas of behavior required of employees in certain situations. Some companies using telephones for customer service support set a standard of the number of rings before the phone must be answered. Standards in performance management might include a certain level of positive rating on "teamwork," or using feedback from an employee's co-workers for measurement. Another example of standards would be limiting the use of e-mail for personal business while at work. Standards and goals are sometimes very similar. A major difference is that goals must be measurable.
3. A *key performance indicator* is a measure of significant activity leading to an important goal. An engineering consulting company might use the number of project proposals submitted to key clients as a measure of activity toward a sales goal.
4. *Models of effective performance* are templates or descriptions of knowledge, skills, attitudes, and expected output required in an area of performance or attached to a particular position. For example, organizations might have detailed descriptions of management styles that they regard as exemplary, or even required, for their managers. A brief case study will clarify this.

Organization: May Co. Inc., One of the Country's Largest Retailers (recently purchased by Federated)

- Products and services:
  - Department store products such as clothing, furniture, etc.
  - Support services such as credit

- Markets:
  - Middle-income families and individuals in suburban centers
  - Lower middle-income families and individuals in some downtown stores
- Structure:
  - More than 200 stores organized into divisions essentially reflecting geographic location
  - Headquarters provided overall direction to the divisions
  - Each division had its own formal leadership
- Primary issues and problems:
  - Customer service was not at the desired level

The organization had committed to quality customer service as an important advantage in the intensely competitive retail industry. They had developed a set of standards or "desired behaviors" for sales associates in their interactions with customers. After rolling out the initial program, including training of the sales associates in good customer service, ratings from customers had improved. But then those ratings flattened out, and customer service was apparently no longer improving.

The customer service executives of May Co. and their performance consultants agreed on a process for improving customer service that focused on leadership within each of the 200 or so stores. Part of the education and training involved developing a model of effective performance, including specific behaviors and policies, for leaders of the customer service process within each store. These managers were provided with a model of effective customer service leadership, a model to emulate. Some six months later, the customer service scores average for all stores showed a very significant improvement.

The leadership of this corporation knew what they wanted and they got it. This is a clear example of "leadership vision."

GOALS, STANDARDS, KEY PERFORMANCE INDICATORS, AND MODELS OF EFFECTIVE PERFORMANCE: MEASURES OF PERFORMANCE

Chapter 5 contains a lengthy discussion of goal setting, KPIs, models of effective performance, and how they form the basis for measurement. Here, the key point is that having these elements throughout the team or organization forms the basis for assessing current performance and deciding whether performance improvement is needed.

A number of points about a system of goals, KPIs, and models of effective performance should be emphasized here:

1. Measurement provides a basis for knowing where performance is and a "baseline" for deciding where it ought to be. This second step is based on the judgment of the leaders of the organization and should be done on a specific "vision" of the company, organization, or department. KPIs help us know where goal attainment is, and whether the activity to reach those goals is sufficient and effective.
2. Knowing the current performance of the team or organization cannot occur without these elements, except at the very general or intuitive level. Having these elements provides the basis for knowing current performance compared to desired performance. The difference between desired and current is a gap or deficiency.
3. Without knowing exactly where performance is, compared to where it ought to be (gap or deficiency), there is little basis for knowing what to improve. This often leads to performance interventions based on fad or program popularity, as discussed in Chapter 1.
4. When there is a clear definition of performance and the current status is known, then there exists a logical basis for deciding what areas need improvement.

IDENTIFYING AREAS OF PERFORMANCE IMPROVEMENT

In the ideal situation, there are shared goals, standards, KPIs, and models of effective performance in each organizational unit, department, team, or sub-group. Performance in these areas is measurable, either directly (e.g., Are we opening new branches at the required rate?), or indirectly (e.g., Is the store manager doing the things in our model that are necessary for building customer service?). This provides the basis for identifying significant gaps and deciding whether or not to engage in performance improvement.

A number of areas are commonly found as the location of performance deficiency. The following provides a list of some of the most common deficiencies:

1. At the top of the list is an unclear or poorly communicated leadership vision for the organization. This deficiency is discussed at the beginning of this chapter and frequently in the following chapters. It is one of the most commonly found deficiencies in organizations of all types, and is the source of many of the deficiencies mentioned below.
2. A small unit or team existing within a certain department, with much of the rest of the organization dependent on the performance of the smaller unit. Examples of this include sales and marketing teams, supervisors in manufacturing, or the IT department in many companies.

3. Poor use of performance management systems — sometimes managers simply do not use the systems. Or, goal setting in the beginning of the performance cycle is not effectively done. Often, there is an unclear connection between performance and promotion or financial rewards.
4. The leadership style is ineffective, or there is a lack of knowledge and skills in managing others, not only at the top level, but also throughout the team or organization.
5. Work processes are poorly defined or ineffective in areas such as planning, hiring, or production.
6. The organizational structure retards coordination and communication between departments, limits production and delivery of products and services, and inhibits customer service.
7. Customer service (patient service in hospitals, student service in universities, etc.) is not a highly valued standard with major emphasis within the organization.

CONCLUSION

Deciding whether performance deficiencies or gaps exist that are significant enough to take action is not simple, even when goals, KPIs, and the other measures indicated above are well established. Thoughtful judgment is required. Often, organizational or team leaders are influenced by their status in the organization as well as their own motivations, personalities, and attitudes. A straightforward process for deciding what, if anything, needs to be done is critical to good decision making. Otherwise, performance improvement efforts can become diminished by organizational politics and egos. In the next chapter we discuss a beneficial process for collecting information and for making decisions about whether and where performance improvement may be needed.

SUGGESTED ACTION STEPS FOR ORGANIZATIONAL OR TEAM LEADERS

1. Organizational or department leaders should ask the following question: Have I (we) defined our strategic vision clearly, including the products and services we are offering and want to continue, markets or customers (external or internal) we serve, and methods of marketing or selling or letting people know what we do?
2. Ask the following question of others in your organization/team: Have we communicated what we are trying to achieve to you and others who are important to performance? Give me the details of

what you understand to be the main products/services we provide internally and externally, and the quality of our service to these internal and external customers.

3. Review the following issues: Where do we have goals, standards, KPIs, and models of performance? Where do we not have these elements? How badly is the omission of these performance measures hurting us? Do we really know how we are performing? In what areas is our performance acceptable, and where is it not up to our standards?

END NOTES

1. The definition of models of effective performance is included in Chapter 3, and examples are contained throughout this book.
2. We discuss the differences and similarities between management and leadership in Chapter 9.
3. Drucker, Peter F., *Management Challenges for the 21st Century,* Harper Business, p. 73.
4. This was done through strategic planning. For a full discussion of the strategic planning process used here and in other cases identified in this book, see Chapter 7.
5. For example, Aldag discusses the relationship between job satisfaction and motivation to performance, while Smither discusses getting members of the organization to "reframe" the desired organizational situation before starting on an intervention. The latter author sees reframing as different from organizational development, an arguable point.
   a. Smither, Robert D., *The Psychology of Work and Human Performance,* Longman, 1998.
   b. Aldag, Ramon J. and Stearns, Timothy M., *Management,* South-Western Publishing Co., 1991.
   c. Kotex, John P., *Leading Change,* Harvard Business School Press, 1996.
   d. Wall, Bob, Solo, Marl R., and Solemn, Robert S., *The Mission Driven Organization,* Prima Publishing, 1999.
6. Perhaps the best treatment of the impact of purpose and goals on human behavior is summarized in: Robbins, Stephen P., *Organizational Behavior, 8th edition,* Prentice Hall, 1998.

Chapter 3

DECIDING ON PERFORMANCE IMPROVEMENT: USEFUL CONCEPTS AND TOOLS

INTRODUCTION

If as discussed in Chapters 1 and 2, the early motivations, attitudes, and decisions of the CEO, president, department head, or team leader regarding performance improvement are critical to success or failure in these endeavors, what, then, can leaders do to increase the chances of success in performance improvement? The first guideline is that leaders must keep an open mind and recognize a crucial reality: *organizational and team leaders are often good sources of information about what is occurring operationally in their organization, but they usually lack the knowledge, experience, and even objectivity necessary for deciding how to improve performance.* A second look at an actual case study discussed in Chapter 1 illustrates this point.

Organization: Matheny Heating and Cooling Service, Inc.

- Products and services:
  - HVAC equipment installation/repair and servicing

- Markets:
  - Residential
  - Commercial
- Primary issues:
  - Declining and insufficient amount of business
  - Located in older neighborhood where the nearby markets were declining

Recall from Chapter 1 that in original discussions with the two brothers who were owners and operators of this company, they believed that customer service was the primary need for improved performance. As discussions continued, it became clear that the owners wanted to improve the company's financial performance and brighten the long-term future. However, that had more to do with redefining their geographic market and rethinking the mix of products and services than with customer service. While excellent customer service is always desirable, in this case providing improved customer service to a market declining because of population shifts would not meet their improvement goals. In fact, as an initial effort with this organization, it would have been a waste of time and money. The owners eventually came to understand the greater potential of strategic redefinitions of products, services, and markets.

It is easy for organizational or team decision makers to start with the wrong motivations and go in the wrong direction for performance improvement. The tangible and intangible costs of those mistakes can be disastrous. Is there a process that can significantly increase the chances of proper motivation and direction for performance improvement? The answer is yes; there is at least one process for defining and managing organizational performance improvement — and probably others as well.

A PROCESS FOR DECIDING ABOUT AND MANAGING PERFORMANCE IMPROVEMENT: IF/WHAT/HOW?

Because leaders and decision makers managing organizations often have limited understanding about how to improve performance in their teams or organizations, help is needed. Even if leaders have experience in developing performance, they benefit from having their perceptions and attitudes reviewed through discussions with others who are well intended and have knowledge about how to diagnose and improve performance. Thus, the role of a performance improvement facilitator or consultant is a valuable one.

Often, organizational leaders have the choice of deciding whether to bring in an outside performance improvement consultant or to depend on internal staff with expertise in assessing and developing performance.

Using carefully chosen outside help means that one may get someone with concepts, models, and experiences not possessed by anyone internally. The biggest limitation to using outside consultation, however, is that outsiders usually do not know the organization thoroughly. In addition, external consultants or process facilitators can be influenced in their judgment by the desire to "sell" more business and generate more income.

Larger organizations (e.g., Anheuser Busch, Inc. (ABI)) have internal performance improvement experts. This internal staff at ABI is frequently used for issues in the various plants and other units of the brewery. However, outside consultants are also frequently called upon, usually because ABI leadership believes that they have experience and knowledge that the company does not have internally.

Whether the decision maker decides to use inside or outside help in the process of deciding about performance improvement, the steps and skills required for success are largely the same. Surprising to many organizational leaders is the fact that while there are some differences in the challenges faced by outside versus inside consultants, the basic objectives in considering performance improvement are the same. Those basic objectives are as follows (1):

- Identification of any areas of the organization where performance may need improvement. Sometimes the needed improvement involves achieving a major positive goal, such as adding a new product line for growth in income. Sometimes the improvement is a more negative problem, such as finding ways to reduce an undesirably high employee turnover rate.
- Identification and definition of performance deficiencies and what, if anything, to do about them.
- Full collaboration and support between the leadership of the organization or team and the performance improvement facilitator (external or internal) to help make performance improvement programs work.
- Evaluation of the performance changes that occur when an improvement program is used, and the development of systems and methods aimed at making the improvement permanent.

During the past decade, starting about 1996, there was increasing emphasis on focusing performance improvement toward achieving improved "business results." (2) While the focus for any form of performance improvement should be on improving organizational results, the need to emphasize this focus leads us to wonder what earlier improvement efforts were trying to achieve. Unfortunately, our experience is that often the goals of performance improvement have been vague or poorly defined.

If organizational leaders choose to look outside their organizations for help in performance improvement, they are often confronted with a wide array of titles and terms that can be confusing. A quick check of the yellow pages in the phone book of any large city demonstrates a wide range of "consultants" focusing on performance. The most generic term is "management consultants," which is used by consulting groups focusing on any area of expertise from finances, to human resources (HR), to technology. Other common terms used by performance consultants include "organizational development," "organizational psychologists," and "HR consulting." The key issue for leaders looking for help is to make sure that any consultant, external or internal, uses an effective process for developing a clear understanding of the performance goals, deficiencies, and opportunities for improvement in the organization.

Many models for deciding on and managing performance improvement exist, whether the area of the organization needing attention is leadership and management, training, technology, or building team performance. One of the most clear and useful models for diagnosing and managing performance improvement (if it is needed) is contained in a graduate-level textbook on organizational development. (3)

THE GENERAL MODEL OF PLANNED CHANGE

Overview

The General Model of Planned Change (GMPC; Figure 3.1) is useful in diagnosing and managing virtually any form of performance improvement: organizational restructuring, training programs, improvements to customer service, etc. The reason for its effectiveness gets to the heart of what is required for *effective* organizational performance improvements. Research and our experience over more than 30 years in this field tells us that there are three critical success factors (CSFs) for successful performance improvement. These CSFs are listed here, starting with what we believe is the most important:

CSF 1: The intense and long-term dedication of the organizational leader to achievement of organizational success and improved performance, where needed

CSF 2: An effective diagnosis of possible areas of performance deficiency and a process for deciding if improvement action is needed

CSF 3: If the decision on CSF 2 (above) is to make performance improvements, then the third most important factor for successful performance improvement is an effectively executed performance intervention

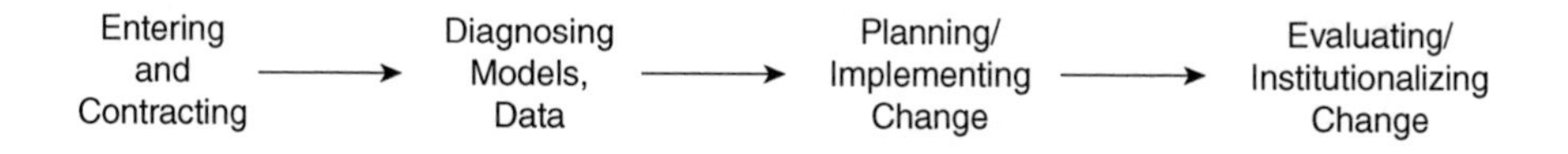

Figure 3.1 General Model of Planned Change (GMPC).

The ineffectiveness of a high percentage of performance improvement efforts derives from failures in CSFs 1 and 2 above. There exists much history regarding interventions and what can work if the leadership of the organization is dedicated, and if the diagnosis of actual performance issues has been done thoroughly. Without leadership dedication, the best facilitation of a well-designed action plan will fail or be of limited success. Without a clear understanding of what issues or barriers to performance must be overcome, the intervention may well be targeting the "wrong" things. The General Model of Planned Change significantly improves the chances that all CSFs for performance enhancement will be met.

The Four Stages

Stage I: Entering and Contracting

The first stage — Entering and Contracting — involves the initial contact between the organizational or team decision maker and the internal or external performance improvement consultant. The key requirement here is that *a trusting and collaborative* relationship is built between the two "sides" in working on the issue at hand. What that means is that the organizational or team decision maker must be candid and thorough in dealing with questions about performance and providing information on deficiencies and issues, even when the realities are unpleasant.

Sometimes, external consultants are desirable because internal performance improvement experts are seen as representatives of corporate leadership, and there is mistrust about the motivation of that leadership. On the other hand, sometimes decision makers do not want outsiders to know about the organization's problems, so they are more open with performance facilitators from inside the company. Assuming that both internal and external consultants being considered by the decision maker possess sufficient knowledge of and skill at performance improvement, the most critical guideline in making the choice of a performance improvement facilitator is finding one who will generate the greatest openness in discussions about performance.

The discussions in the entering and contracting stage focus on the following objectives:

- Identification of areas in the organization where performance improvement may be needed. Again, the focus can be on improvements for major steps forward in overall organizational success, or on "stopping the bleeding" in some negative occurrences.
- An initial broad definition of the identified performance deficiencies and who and what factors are involved. Here, the focus is not on finding fault, but on identifying causes of any performance deficiencies.
- A discussion regarding how to proceed with further inquiry into the issues of performance deficiency. Here, tactical questions about methods of additional data collection are also answered.
- An introductory discussion of the role of the organizational or team leadership in supporting the diagnosis that follows this first Entering and Contracting stage.

The organizational or team leader should maintain candor in communication and recognize his or her critical role in the entire performance improvement process from beginning to end and beyond. The performance improvement facilitator or consultant has to establish and maintain commitment to using independent judgment in what are often sensitive discussions. That sets the stage for the consultant to add value throughout the entire process, no matter how far it goes. It is sometimes difficult to remember that, although the decision makers lead the teams or organizations, they do not necessarily completely understand the issues, causes, or cures for performance improvement. The facilitator provides the best value by asking insightful questions, proposing alternative ideas and approaches, and disagreeing with the leadership when there is a basis for doing so.

Stage II: Diagnosis

The second stage of the GMPC — Diagnosis — builds directly upon the first stage. If an effective and collaborative relationship is built between organizational or team leaders and performance improvement consultants, then Diagnosis has a good chance of working well. Our experience shows that organizational leaders quickly form impressions about consultants and whether they will be useful in performance improvement processes. This "first impression" about their usefulness is usually formed quickly whether the possible consultant is from inside or outside the organization.

Diagnosis is both the most difficult and the most important of the stages in the General Model of Planned Change. It is the most important because if the diagnosis is not accurate, you may end up working on the wrong performance deficiencies and causes. Diagnosis is aimed at answering the following questions:

- What are we trying to achieve with this organization or team? What is the strategic vision, and is it shared broadly throughout the organization? What are the shared goals or objectives and standards of the organization, and are they connected to the business vision and strategy?
- What areas of deficiency or needed improvement exist? The organization cannot do everything at once, so what should the initial focus be?
- What is the nature and what are the causes of performance gaps or deficiencies?
- If we decide to get involved in planning and implementing change, what intervention is most apt to improve the performance?

Words such as "deficiencies," "problems," and "issues" are often worrisome or confusing to organizational leaders. This is particularly true if their organization is apparently successful. They sometimes seem to believe that their organization or team is perfect, or at least has little room for improvement. The question that the diagnosis stage answers is: Do we have room for significant improvement in performance? No matter how good we are now, can we do significantly better? (4)

Sometimes, organizational or team decision makers ignore deficiencies and problems for so long that their unit gets into serious performance trouble. This often means that it is too late to make changes resulting in major performance improvement. Diagnosing performance on the basis of crisis can lead to premature judgments about what is really going on and to rash judgments that "anything will make it better." The concept of "continuous improvement," made popular by quality programs, should be a guiding principle for organizational or team leadership. "We can always get better" is a perspective all leaders, no matter what their organization, would do well to live by.

Effective diagnosis requires three things: (1) a model of effective performance, (2) measurement using data to compare the organization or team being reviewed to the model of effective performance, and (3) providing feedback to the organizational or team decision makers.

A model(s) of effective performance is needed for the area of deficiency being reviewed. That means that the persons doing the diagnosis must know what organizational or team performance of the type being scrutinized looks like when it is "excellent." Different concepts and approaches provide examples of effective performance. Some examples include:

- What is the vision for this team at its best?
- What is the model of excellence?
- What are the benchmarks for best in class?
- What are the major standards for top performance?
- What goals or objectives are we trying to achieve?

All these questions try to get at the same thing: what does top performance in this area look like? Detailed examples of useful models of effective performance are presented later in this chapter and throughout this book.

Measurement using data to compare the organization or team being reviewed to the model of effective performance is the second requirement for making diagnosis work. Data can be obtained through interviews, surveys, observation, or existing information collected by the unit during its normal operations ("intrinsic data"). Which source to use is determined by practical considerations such as the area of concern and the size of the group or team involved. For example, if morale and turnover in a very large department is the concern, then a survey may be the most efficient data source because of the ability to cover a large group of people at a reasonable cost. If the leadership style of a small group of executives is the focus, then structured feedback from their direct reports and one-on-one interviews may be the best method of data collection because of the ability to probe deeper into individual style.

Many of today's organizations collect all kinds of data connected to performance of teams and individuals. Data collection without some idea of what factors are crucial to effective performance, however, is often random and wasteful, leading to few significant changes. Too many organizations, for example, do surveys of employee morale or organizational culture, then do very little to improve anything. When this happens, frustration among those completing the surveys grows, and performance improvement efforts receive a negative reputation.

An example of an efficient survey, the Organizational Climate Survey (OCS), based on a model of effective organizational performance will help clarify this point.

Dimensions for the OCS include:

1. Communication: The general degree to which communication is open, based on mutual trust and respect, clear, and effective in contributing to the achievement of organizational goals.
2. Responsibility: Members of the organization are given responsibility to achieve their part of the organization's goals. The degree to which members feel that they can make decisions and solve problems without checking with superiors each step of the way.

These first two questions of the survey are scored by having respondents indicate on a ten-point scale where they think their organization or team actually is on the dimension. Respondents also indicate on the same scale where they desire the organization or team to be. Group averages are calculated for both *actual* and *desired*, and the gap between the two averages is then reviewed and discussed.

The key point is that these two questions reflect a model of effective performance that includes the following principle: organizational goals are central to the performance of the organization or team, and it is important to review communication and the responsibility given to members of the organization in light of their contribution to the achievement of those goals.

To continue with this illustration, data collection about communication in an organization, team, or department is one of the most popular survey topics. The notion of "effective communication" becomes more focused when talking about communication pertaining to the achievement of goals. The same can be said for "responsibility," or any other area of data collection.

Feedback to managers, sometimes called "360 degree feedback," has been very popular in recent years. But unless those collecting and reviewing the data have a clear model of effective managerial performance, the data is difficult to interpret, and usually the interpretation comes down to the attitudes of the persons doing the interpretation. That is the essence of what is meant by "subjectivity."

The concept of *models of effective performance* is based on the belief that we have principles and concepts that describe *better* or even *best* known performance for organizations and teams, and for within specific areas of responsibility (e.g., good teamwork, effective selling skills, or good supervisory skills). It is also based on the belief that we can identify a model of effective performance for positions at the individual level (e.g., sales manager, CEO, etc).

This does not mean that there is one "right" organization, a single most desirable organizational structure, or final answers about hierarchy in structure versus teams or matrix structures. The nature of the products and services and the markets served, among other things such as size of the organization, all influence which structure will work better, and which ones will not work at all. But as Drucker has noted, while there is no one "best" organization, "[t]here are some 'principles' of organization to be followed." (5)

The concept of *models of effective performance* also does not mean that there is one right way to manage or lead people, with no consideration of their differences in personality, motivation, knowledge, and skills. In fact, there is evidence, as evidenced in Chapter 9, that the most effective managerial performance involves managing with consideration of the individual differences between those being managed. Similarly, selling must be adjusted to the nature of the business. Selling pet food to thousands of veterinarians across the country requires prospecting, whereas serving a customer who comes into the drugstore does not. Still, there are guiding principles for all kinds of effective selling in these and other situations. An example is the importance and use of skills necessary for identifying customer needs and wants, and matching products and services to those customer needs.

The third element for productive diagnosis is *providing feedback to the organizational or team decision makers* comparing the current situation to the model of effective performance.

This discussion began by stating that diagnosis meant knowing what effective performance looks like in the area being reviewed and then collecting data about current performance in relationship to that definition of effective performance.

Obviously, providing feedback data to organizational leaders is a critical decision point. The gap or deficiency between the existing situation and the model of effective performance may be small enough to leave things alone. If the gap is significant, some *planning and intervention* is required.

During diagnosis, the roles and responsibilities of the organizational or team decision maker and the performance improvement consultant are both critical and yet somewhat different. The consultant or facilitator keeps the process moving, contributes models of *effective performance,* collects and presents the data, and makes recommendations. The decision maker provides data for the diagnosis, supports additional data collection as needed, and ultimately makes the decision about whether planning and intervention should be done. This is a collaboration that is absolutely necessary for performance improvement to work.

A brief case of excellence in diagnosis during a highly tense period of time will clarify the discussion above.

Organizational Unit: A Large Information Technology Department

- Products and services:
  - Software and hardware purchases and installations for their company
  - Desktop support
  - Maintaining Internet functions for their company
- Markets:
  - Internal and external customers
- Structure:
  - Large department with a vice president for IT, four top department managers, and 60 operators

- Primary issues:
  - Indications that IT support was falling behind the needs of the company

Leadership of the business side of this company had an impression that the IT department was not doing what was needed to keep the Internet functioning and the technology up-to-date. This was crucial to the firm because a high percentage of the communication between the company and its customers occurred over the Internet. The "business interests" of the firm were not being fully met by the IT department.

Leadership of the company commissioned an audit of the IT function from an external technology consulting firm. They also decided to look at the functioning of the department from a leadership and organizational climate perspective. They used the Organizational Climate Survey (OCS) described above to gather the data for this area.

The results from both the technical and the leadership and organizational climate perspective were remarkably similar. The technology audit indicated that there were a number of areas where the technology was in need of dramatic improvement and updates. The OCS survey indicated a series of mistakes being made by the IT department's management that resulted in ineffective communication between different organizational units within IT, and between managers and the IT operators. The deficiencies in the technical and "human" sides of IT reinforced each other. Appropriate interventions, both technical and organizational, were initiated by the firm's leadership to improve the situation.

Stage III: Planning and Intervention

The third stage of the Model of Planned Change — planning and intervention — only occurs if the diagnosis identified significant problems or deficiencies in the organization or team. In that case, the interventions identified in the diagnosis stage are extended and carried into action. This is not to say that additional interventions might not be added as the intervention is planned or even as it is carried out. Another brief case study here will clarify the point about discovering new issues after interventions have begun.

Organization: Motor Appliance Corporation

- Products and services:
  - Customized electric motors
  - Commercial and industrial battery chargers

- Markets:
  - Original equipment manufacturers such as golf cart manufacturers, etc.
  - Users such as railroads, etc.
- Structure:
  - Corporate office in St. Louis
  - Two plants, one in Missouri and one in Arkansas
- Primary issues and problems identified during diagnosis:
  - Managers lack sufficient knowledge and skills about managing

Discussions with the new CEO of this organization involved questions about strategic versus operational needs. Both performance improvement areas had deficiencies. The decision was to develop the top management team.

The management development program began, but some resistance to the program from selected managers existed from the beginning. It soon became apparent that there was an additional problem affecting performance that had not been uncovered during diagnosis. Specifically, there were major conflicts occurring between the sales force for one of the product lines and the production people located in a plant that manufactured that product line. Production believed salespeople were making promises to the customer that production could not keep. Sales charged production with not working hard enough to meet deadlines and design quality specifications. In this case, the overall management development intervention was expanded to include work on inter-team communication and decision making.

Lists of potential interventions for performance improvement will vary, depending on who is providing the list. The interventions included below focus on the human side of the organization or team, rather than strictly on physical equipment, finances, or information technology. But money, technology, and equipment are not excluded in these areas. The communication equipment and computer systems used by the organization or team, for example, will probably be part of a strategic plan, but the strategic intervention is still primarily about people choosing and then using those support systems strategically. Effective selling and customer service will be financially beneficial but these programs are largely matters of the performance of employees.

The following is a list of *major interventions* for performance improvement (6):

1. Strategic planning, cascading goals, and KPIs
2. Improving the hiring process
3. Restructuring to enhance process or process improvement

4. Improving effective teamwork for any team in the organization and building inter-team cooperation
5. Developing intact teams in subject area: management or leadership teams, sales teams, customer service teams, etc.
6. Performance management
7. Interventions aimed at improving organizational effectiveness and efficiency
8. Methods aimed at developing individual performance
9. Training, development, and education

Stage IV: Evaluation

Stage IV of the Model of Planned Change is evaluation. Evaluation of changes in organizational, team, and individual performance is frequently given little attention in actual cases of intervention. As often as not, this inattention to evaluation is based on the belief that one cannot measure "the soft" skills such as how well managers manage or team members function. This problem with measurement starts with insufficient data collection during diagnosis and not having a model of effective performance. If the diagnosis stage is incomplete in data or models of effective performance, then measurement of improvement is usually just a measure of "contentment" with the intervention.

Experts in training and development, most notably Donald Kirkpatrick, have been leaders in advancing evaluation of performance improvement. Many of the approaches and techniques useful in measuring performance improvement from training interventions are also useful, with some adjustments, in evaluating performance improvement from other performance improvement interventions. The key is focusing on measurement of specific "outcomes." For example, have we accomplished the strategic objectives developed in our strategic plan? Have our efforts at improving the hiring process led to a reduction in turnover, or better performance by new hires compared to earlier hires?

Evaluation of performance improvement efforts is important for a number of reasons. First, one must know what, if anything, about the intervention produced improvement so one can get started on finding ways to keep or "institutionalize" what has been effective. Second, maintaining the motivation and efforts of those involved in the intervention is more likely to occur when they can see "real" progress shown by evaluation.

The third benefit of evaluation for performance improvement is in many ways the most important benefit. As suggested previously, and as will be discussed more fully in the chapter on strategic planning, organizations tend to either constantly work to improve or they tend to decline.

In fact, the inherent nature of organizations is to grow, mature, and decline. We have known for a long time that this is the natural trend in products and services. We now understand that it is also a natural tendency of entire organizations. This is especially true in today's "one world economy" and its related growth in competition for organizations.

This is a good place for a related side note directed toward those reading this book who lead or manage not-for-profit organizations such as government units, universities, or some healthcare organizations. The reality is that all of these organizations experience competition in some form or another. Clearly, most healthcare organizations compete for patients, and universities compete for students and the financial support of foundations and wealthy citizens. In this era of "privatization" of some government services, even local and state agencies face competition. Most significantly, no organization will do well without the support and approval of other groups or agencies. Healthcare organizations must maintain licensure, universities, in most cases, need to maintain their accreditation, and government agencies need to maintain some support from the other agencies that fund and regulate them.

To maintain their funding and avoid problems with their stakeholders, these organizations must maintain sufficient performance to avoid alienating those stakeholders. In a very fundamental sense, all organizations have their "clients."

Measurement of improved performance of individuals, teams, and organizations is not as "objective" as calculating profit at the end of the fiscal year. However, if we know specifically what the organization or team is trying to accomplish, then measurement and evaluation of performance improvement efforts is somewhat more "objective." This means that it is possible to apply metrics to significant areas of performance, collect data, and get some agreement on whether things have improved.

In his initial work, Kirkpatrick identified four levels of evaluation, although as mentioned above, he described them largely in terms of their application to training programs. (7) The four levels described are:

1. Evaluating reaction: how people felt about the intervention
2. Evaluating learning
3. Evaluating behavior
4. Evaluating results or performance

Part II of this book describes a number of performance improvement interventions in detail. It shows how this basic Kirkpatrick model, with some adjustments, applies to evaluating other types of intervention in addition to training and development.

Knowing what the organization or team should be trying to accomplish starts with a clearly defined strategic vision. In fact, as stated in a previous chapter, the strategic vision provided by the leadership is the basis for defining performance and then deciding if improvement is needed. (8) If a department's or team's efforts are determined by the nature and direction of the overall organization (such as a production department in a factory), then this strategy of the entire organization should be the focus for defining performance of all the smaller organizational units. If, on the other hand, the team or smaller unit has a lot of freedom in deciding what products or services it will provide, internally or externally, then that team may well need its own strategy. This "department strategy," of course, should be developed in support of the strategy of the overall organization. A marketing or selling team, for example, with lots of choices to make about what they are going to sell to which customer group, may benefit from their own strategy. Similarly, a large human resources department, with plenty of choices to be made on HR products and services to be offered to various groups inside the organization, may also need a strategy for its own direction consistent with that of the corporation.

A MODEL OF EFFECTIVE PERFORMANCE: THE ORGANIZATIONAL SUCCESS MODEL (9)

As previously discussed, the diagnosis stage is both the most important and the most difficult stage in the General Model of Planned Change. The difficulty with diagnosis is often due to the lack of agreement on *models of effective performance.* The literature about managing or leading organizations is full of such models. But one book or speaker will recommend one version of, for example, successful organizations or the "best" way to provide leadership. If one looks on the next shelf in the bookstore, however, one will find different versions of how to build your successful organization or be an effective manager. Ultimately, the leadership of each organization must decide what models of effective performance they believe to be the best for their organizational situation and stick with them.

The following model of effective performance (Figure 3.2) is frequently useful in helping organizations or teams decide what performance interventions, if any, they need to undertake.

Organizational Success = Clear and Shared Strategy +
Effective and Efficient Operations

The best use of this model is to begin with a review of its first principle, that of *organizational success coming from a clear and shared strategy,*

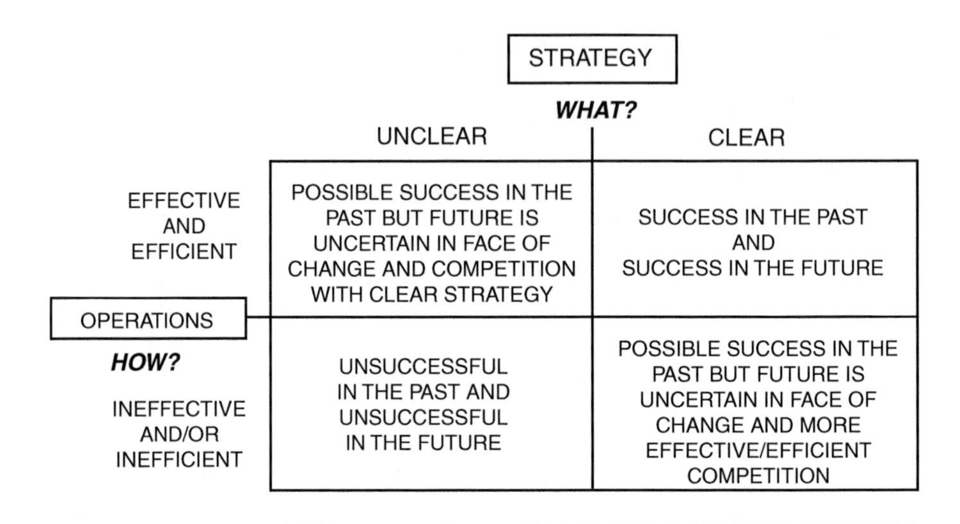

Figure 3.2 Model for organizational success.

and effective and efficient operations. Strategy is comprised of "what products and services, marketing strategy," and other supportive elements the organization/team is striving to obtain. Operations is "how they manage, hire, provide customer service, reward and recognize, etc."

Diagnosis occurs using this model by discussing whether the strategy is clear and shared. A strategic concept statement, which is really a general marketing plan, may be unclear for a number of reasons. These reasons can include new competitive activity, outdated products and services, shifts in the use of disposable income of the organization's market, or overall decline of its market. (10) McDonnell Douglas Corporation, years before being purchased by Boeing, began to experience major strategic problems when the Cold War ended and the U.S. military reduced its budget for war machines. This international reality affected a large number of defense contractors and their suppliers.

One of the major deficiencies in clarity of the strategic vision, however, is that it may exist only in the mind of the organizational or team leader and not be known by others in the organization who are needed to make it happen. But whatever the reason for the strategic vision being unclear or unshared in the organization, this situation has negative consequences. If the strategy is not clear or shared by at least the top group in the organization or team, there is a lack of common direction, usually resulting in a lack of coordination. If this situation exists, a performance improvement intervention is necessary.

A brief case study will be useful for clarification.

Organization: Ehrhardt Tool and Machine Company

- Products:
  - Precision tools and dies
  - Specialized equipment
- Markets:
  - Nuclear
  - Appliance manufacturers
  - Furnace manufacturers
  - A number of other market segments
- Structure:
  - New CEO
  - Four other top managers
  - Owned by a larger company
- Primary problems and issues:
  - New CEO and other managers felt the need for change in how they managed the business.

The CEO of this organization began discussions with us regarding the need for agreement on the future direction of this company. We decided to use the Organizational Success Model as the focus of a discussion with all the top managers. This agreement developed during the discussion:

1. The managers agreed that they had some operational issues and the need for a more productive selling effort.
2. They wanted improved morale.
3. The managers needed to learn additional management skills.
4. The management team needed to learn to work together more closely.

However, the fundamental issue put to this management team was where to start their performance improvement efforts: with improving operations or with creating and operating a strategic plan. They decided to start by creating a strategic plan, using the rationale that the other operational changes needed to be directed toward a clear strategy. At the time of this meeting with the managers, that clear and shared strategy did not exist.

There are many designs for a strategic plan. Some of them include a mission statement, and a list of values. (11) We have found, however, that at least the following elements of a strategic plan are crucial to performance improvement:

- A strategic concept statement: a detailed statement of what products and services, markets, methods of marketing/selling, and distribution system (among other things) the organizational/team leadership desires and intends to have in place at some point in the future. Again, this is usually referred to as the "strategic vision," and is thoroughly detailed.
- A detailed analysis of the current situation within the organization or team in its significant surroundings in light of where the organizational decision makers want to go (the strengths, weaknesses, opportunities, and threats — or SWOT analysis).
- A set of specific and immediate strategic objectives for taking the first steps toward the strategic vision.

Remember that the Strategic Plan is a document that provides immediate and future direction for an organization or, where appropriate, for a team. It creates the insight and context for diagnosis of operational issues. It improves the ability to make use of measurement in all areas of the organization. The Strategic Plan includes strategic objectives and key performance indicators (KPIs), and establishes the basis for measurable goals throughout the organization. The strategic objectives should be specific and measurable so that there can be evaluation of the progress of the top leaders of the organization or team. Some examples of specific measurement include development of a new product line for our fabrication division by June 30th; add thirty new branches within six months; or deliver sales training aimed at increasing business 20 percent by the end of the year.

We discuss strategic planning as an intervention process in Chapter 6. Here the primary point is that strategic planning at the top level of the organization, itself a performance improvement process, can set the basis for other performance improvement interventions because we begin to know what performance best serves the strategy. The strategy becomes the basis for immediate and future action, including, but not limited to, additional performance improvement interventions.

During the diagnosis stage, it may become obvious that the strategy is clear enough for the moment, and that one or more operational areas needs performance improvement instead. Not all organizations need to start their performance improvement efforts with a strategy. But some form of evaluation of current performance and vision of what performance should be must exist before any reasonable performance improvement can occur.

While the graphic above describing the stages of the General Model of Planned Change looks like the stages begin and end, diagnosis can actually occur at any time. In the Ehrhardt case described on the previous page, as the strategy unfolded, diagnosis indicated even more clearly that there was a need to improve the performance of the sales team. At Ehrhardt, the data collection during the initial diagnosis provided the basis for strategic planning and the strategy led to further efforts at improvement in the sales team's results. There is a discussion of a model of effective performance for selling in subsequent chapters.

Whatever the performance improvement intervention, strategic planning or enhancing performance development in an operational area, measurement is important. Again, the important phrase here is, "if you can't measure it, you can't manage it." Related to that is, "if you can't measure it, you can't improve it." Clear strategy, key performance indicators, and goals throughout the organization are the basis for measurement, performance management, and, where needed, performance improvement.

SUGGESTED ACTION STEPS

1. Commit to constructing a performance assessment and improvement team. Decide who should be involved in discussing organizational performance. Who should be on this team from one's leadership and operations people? Who should be the performance improvement facilitator or consultant?
2. Decide how to maximize candor in performance discussions.
3. Decide how and when to start the initial discussion with the newly formed performance improvement team (entering and contracting). Structure these meetings enough where there is a brief agenda and keep notes to keep you on track.
4. Decide what diagnosis is needed. Are the issues macro or micro (the entire organization or a limited number of units)? What kind of data does one have? What kind of data should be collected?
5. What is the model (or models) of effective performance here? What are the standards, goals, or performance objectives? If there are none, how does one start to develop them?
6. If there is a significant area of deficiency, what performance improvement intervention does one want to undertake? What are the costs and risks?
7. If one decides to proceed with the intervention, what type of evaluation should be conducted? How can one maximize chances to know how and where the intervention helped?

END NOTES

1. Part III discusses stabilizing or "institutionalizing" performance improvement once the performance improvement has been initiated and progress has been defined by evaluation.

2. Dubois, David D., *Competency-Based Performance Improvement: A Strategy for Organizational Change,* HRD Press, 1993; Robinson, Dana Gaines and Robinson, James C., *Performance Consulting,* Berrett-Koehler Publishers, 1996.
3. Cummings, Thomas G. and Worley, Christopher G., *Organizational Development & Change,* 6th edition, South-Western College Publishing, 1997, p. 32. See the later editions, particularly the 8th edition, 2005, for significant additional insight into performance improvement systems.
4. A problem is a barrier to a goal or objective. Therefore, diagnosis must start with goals or directions of the organization or team, and then get to the barriers or deficiencies. The concept of deficiencies in performance, sometimes called "gap analysis," is most clearly developed in diagnosis for training and development, but applies to other performance development needs as well. Discussion of this application to performance improvement generally is contained in Chapter 2. See: Blanchard, P. Nick and Thacker, James W., *Effective Training,* Prentice Hall, Inc., 1999, Chap. 4; and two books by Dana Gaines Robinson and James C. Robinson, *Training for Impact,* Josey Bass Management Series, 1989, and *Performance Consulting,* Berrett-Koehler Publishers, 1995.
5. Drucker, Peter F., *Management Challenges for the 21st Century,* Harper Business, 1999, p. 13.
6. For a useful distinction between training, development and education, see Dubois, David D. Ph.D., *Competency Based Performance Improvement,* HRD Press, 1993, p. 4.
7. Kirkpatrick, Donald, *Evaluating Training Programs,* Berrett–Koehler Publishers, Inc., 1998.
8. Kotter, John P., *Leading Change,* Harvard Business School Press, 1996, Chapters 2 and 5 in particular.
9. Credit for developing this model, along with a number of others in this book, goes primarily to W.C. "Mike" Weaver, President of Achievement Associates, Inc.; Achievement Associates, Inc., also gives credit to Benjamin B. Tregoe and John W. Zimmerman, *Top Management Strategy,* Simon and Schuster, 1980.
10. Drucker, Peter F., *Management Challenges for the 21st Century,* Harper Business, 1999 Chap. 2, p. 58.
11. Wall, Bob, Sobol, Marl R., and Solum, Robert S., *The Mission Driven Organization,* Prima Publishing, 1992, 1999; see Chap. 3.

Chapter 4

IMPROVING PERFORMANCE FOR NEW ORGANIZATIONS

WHY THIS CHAPTER?

The first edition of this book, published in 2000, did not include a chapter on performance in new organizations. That was partly an oversight in planning the original edition. However, the omission was also due to a lack of appreciation on our part of the continued volume in our economy of both new "start-up" organizations, and mergers and acquisitions.

The selection of what will be treated in this chapter as "new organizations" is guided by whether the organization has new leadership, a new business focus, or new structural changes. These changes often lead to major change initiatives aimed either at enhancing or minimizing disruptions in the performance of the organization.

The common denominator of the organizations in the following list is that they are all experiencing major change, which usually results in dramatic concern for performance (again, either for increasing performance or for minimizing disruption). The types of organizations discussed in this chapter include:

1. New start-up organizations with little or no previous history of provided products or services to clients, patients, or students. A very substantial percentage of the U.S. economy is comprised of organizations having been in business for less than five years.
2. A new organization comprised of the combination of two or more previously separate organizations. This is usually due to one organization buying another, or, less often, the result of a more or less friendly combination of organizations previously operating independently.

3. An organization in which there is a very significant change in the top leadership due to the addition or loss of an important leader. A clear example, one used in an actual case later in this chapter, is a sole proprietorship that adds one or more additional owners, becoming a partnership or perhaps a corporation.
4. A new department or unit within a larger organization, when the new unit will be permanent.

SCENARIO 1: THE NEW START-UP ORGANIZATION

When a person (or perhaps two or three people) decides to start a new business, a number of questions must be answered. Some of these questions are unique to a new organization, while others will be very similar to those confronting organizations that have been around for some time.

1. What (products and services) do we intend to provide that will be valued by others? Are these products or services that others value something that I (we) as an organizational leader(s), also value? That is, will my conviction about the value of what we provide keep me motivated?

   Many books and articles have been written about being an "entrepreneur," which is a term often used to mean the owner and founder of a small business. Many universities are now providing courses and even degrees in entrepreneurship. The key characteristic of successful start-up business entrepreneurs is dedication and belief in what they are providing. The belief of the entrepreneurs not only keeps the owner-entrepreneurs motivated, but also often helps them in getting customers. Enthusiasm sells.
2. Who should we target as those who will value our products and services and pay to have them (markets)?
3. How do we get needed information about our products and services to our markets so as to have them obtain what we provide (marketing and sales)?
4. What type of legal organization do we want to use (sole proprietorship, partnership, S corporation, C corporation, etc.)?
5. What type of organizational structure do we want to have (roles and responsibilities)?

The first three questions listed for this scenario are the very basics of what was referred to in previous chapters as the "leader's vision." With a new organization or business, the basics are an important start. Elaboration into a full strategic plan comes later with experience, successes, and some failures.

A very high percentage, probably greater than 50 percent, of new businesses close or go bankrupt within the first 5 years of their existence. While the percentage is a matter of debate, it is also true that a significant percentage of personal bankruptcies have a business involvement as part of the financial cause. (1) It is our perspective that one of the major causes of failure for new businesses is that they do not define what they want to provide and to whom they want to provide it. They fail because they try to be too much to too many people and do not clearly define what business they are in.

Once the basics for the new organization are set through clear answers to the above questions, other business decisions must follow just like they do with organizations that have been in existence for a good deal of time: What kind of people do we want to hire, and how many? What suppliers do we need, and in which areas? Etc., etc.

It is never too early for a new organization to begin developing baseline numbers to measure its current performance and to later decide what areas of performance need improvement. Success in starting a new organization requires a combination of what appear to be contradictory personal characteristics:

- The creativity and innovative spirit necessary to visualize a new business
- The systematic and persistent dedication to do all the things, including what is listed here, to make this "vision" happen

This combination of characteristics in organizational leaders is somewhat unusual but important, particularly with a new organization. Some have wondered whether or not "entrepreneurship" can really be taught. (2) Whether these characteristics can be learned is an interesting and significant issue for those wanting to start a business. But, whether learnable or not, some way of including the creativity and visionary ability with the ability to deal with daily systematic operations is required. One start-up company we spent time working with seems to have found an answer.

Organization: A Healthcare Services and Products Firm

- Products:
  - Sleep issues analysis and treatments
  - Products related to sleep issues: medicines, oxygen, masks, etc.
- Markets:
  - People with diagnosed sleep issues: narcolepsy, shaking leg syndrome, etc.
  - Rural areas in three states

- Structure:
  - Owned by two partners, one managing the sleep services unit and the other managing the products unit
- Primary issues:
  - Competition from hospital-based programs
  - Working relationship between the two partners

Both of the partners were very intelligent and personally admired each other. One of the partners was a visionary regarding the products and future of the organization. She was highly creative and charismatic, and others tended to follow her.

The other partner was equally intelligent but more concerned than her partner in achieving organizational stability, getting the bills paid, and setting up systems and procedures.

Together they had what was required to be successful with their new enterprise. But working together successfully required consistent attention to the major differences in how they approached leading the organization. Creating a strategy with common objectives was helpful in their collaboration.

SCENARIO 2: MERGERS AND ACQUISITIONS

This type of "new organization" is different from the first one because it combines previously existing organizations that have history, organizational structure, and human resources. A merger is a combination of two or more organizations into a new form of organization. An acquisition is one organization buying another one. In the latter case, the organization bought often largely or entirely ceases to exist.

We noted previously that start-up organizations have a high failure rate. Mergers and acquisitions also have a questionable rate of success. (3) For simplification, it is useful to think of the problems faced by this form of new organization as one of two types: (1) those difficulties that occur because the combination was a bad idea in the first place; and (2) those problems that come from cultural conflict between the two organizations.

The problems that come from a bad combination have to do with the reasons why one or more of the participating organizations wanted the combination. One set of reasons for acquisitions or mergers involves the desire to add new markets or products by combining with or acquiring a business in operation. For example, a large food distribution company may choose to purchase a food manufacturing company, thus providing them both with additional products to sell and with cost cutting from producing and distributing their own product rather than buying food or distribution services from an outside source.

Sometimes an acquisition is motivated primarily by the acquiring organization wanting to buy customers or business from the acquired organization. So, a larger bank may buy smaller banks to get their customers and the associated accounts. Another example is when private schools buy other schools to get their existing educational programs and student lists.

The primary point here is that sometimes the combination looks more attractive financially than it actually turns out to be. This is why "due diligence" is so important when mergers and acquisitions are in the early stages.

The human side of the problems that occur with mergers and acquisitions is the major focus of performance improvement programs. Whatever the wisdom of the initial decision to combine two or more organizations, once the combination is underway, there are a number of actions that can help the leadership minimize performance disruption and set the basis for performance improvement. Most of these actions have to do with the treatment of and communication with people in these organizations. Briefly stated:

1. Leaders of the changes in mergers and acquisitions should make use of a simple reality of human communication in organizations: if people are not told what is going on, they will create their own answers, correctly or incorrectly. Within limits, it is better for leadership to communicate openly rather than let the rumor mill define what is occurring.
2. Leaders of these changes should make sure that other managers support the changes that are going on and do so with employees.
3. Communication between leaders and employees should be frequent, especially in the early stages of change in the organization. It should also be two-way, with employees able to ask questions. The so-called "town hall" approach to communication is a useful model.

Clear communication between and from leadership of the new organization formed from a merger or acquisition will not solve all problems — nothing solves all problems. However, clear communication will help reduce those problems that come from negative morale, confusion, and anxiety.

During the early stages of the new organization, leadership should clarify its vision for the organization during the communication just discussed. What will be provided by the organization, and will certain products or services be changed? Who is responsible for what? Will the major markets or customer groups change soon?

There are two primary issues that affect employee performance during early stages of merger or acquisition reorganization. The most basic is the question of who will lose their jobs and who will be kept. When possible, it is smart for the leadership to at least clarify the immediate situation regarding any changes in the workforce, including any steps being taken to decide on future cuts. Again, if this issue is not discussed, employees will access the rumor mill, often concluding the worst, and major difficulties will occur.

A second major issue, especially significant with acquisitions, is the degree to which people in the acquired organization will resist new systems, methods, and procedures. Personalities vary in their willingness to make changes in how they do things, but the employees' ability and willingness to change what they do are higher when they know why the changes have been made. Leadership has the right to decide how things will be done but will get the greatest amount of willing adjustment to new ways when employees see why the change should occur. This is true even if employees do not agree with the rationale for the change. People like to "be in on things," especially during stressful times of change.

SCENARIO 3: SIGNIFICANT CHANGE IN THE LEADERSHIP OF AN EXISTING ORGANIZATION DUE TO THE ADDITION OR LOSS OF SIGNIFICANT LEADERS

This type of new organization emerges in different ways. Either a highly influential leader or owner has left the organization, or a new partner/owner, perhaps more than one, has joined an existing leadership. The case of Landshire Company, as discussed throughout this book, provides a clarifying example.

Organization: Landshire Sandwiches Inc.

- Products:
  - Sandwiches
  - Other items, including tamales, roller-grill items, etc.
- Markets:
  - Convenience stores and other food providers in 16 states
- Structure:
  - Family owned
  - CEO, small group of top managers, large selling staff, including route drivers
- Primary issue:
  - Considering additional large distribution channel for its products

Landshire Inc., founded many years ago and initially run out of the founder's garage, had grown to a multi-million dollar company. Approximately five years earlier, the company initiated strategic planning and had revised the strategic plan each year as needed. Through strategic thinking and some creative selling, the company decided that providing its product to the entire United States was possible if they used vending companies as their market. Vending companies would buy the products from Landshire and sell those products to their customers. This would allow Landshire to essentially "go national" without having route drivers and trucks outside the 16 states where they did direct delivery.

At about the same time, the son of the founder was ready for a top-level position. The CEO, not a family member, was a valued and experienced executive the company did not want to lose. As is always the case when leadership personnel expands, the need was to define which of the two top executives would do what. Solution: the CEO would maintain primary responsibility for the route business, which he knew very well. The founder's son, who was designated as President, became responsible for the national business.

This expansion of the company's business and the creation of a second top-level manager fit well together. But what if there is no commitment to create a "new" area of responsibility while adding a top manager? The guidelines are the same whether or not there is a new area of business for the new executive. The need in both cases is a clear definition of roles and responsibilities for each executive. It is just a little tougher when the new leader's area of responsibilities must come from the original leader's previous areas of responsibilities.

SCENARIO 4: A NEW DEPARTMENT OR UNIT WITHIN A LARGER ORGANIZATION WHEN THAT UNIT WILL BE PERMANENT (I.E., NOT A TEMPORARY PROJECT TEAM)

Here, an existing organization has decided to create a new unit that has not existed before. This new unit could be an entirely new division with its own customers, president, financial responsibilities, etc. It could also be a department within an organization where growth or change in the larger organization has led to the creation of the new department with identifiable responsibilities. An example of the latter situation might be the creation of a "product development" department where new products had previously been the responsibility of the marketing department.

The performance improvement issues here are largely the same as in Scenario 3. The primary need is to establish a clear initial definition of roles and responsibilities between the new organizational unit and the larger preexisting organization. In fact, in all four types of new organizations discussed in this chapter, at some point the urgent need becomes that of defining roles and responsibilities between leaders and organizational units. The differences between the four scenarios have to do with when and how to define who does what.

It might seem that reminding readers of the need to define roles and responsibilities is to point out the obvious. Unfortunately, a lack of clear objectives or goals and clear definition of who is responsible for various areas of performance is often a primary cause of poor performance. As frequently mentioned in previous chapters of this book, performance can only be measured if the goals are clearly defined. However, people within the organization must also own up to their areas of responsibilities for achieving organizational goals. When they do not take responsibility, it is sometimes because the goals were poorly defined. At other times, avoiding responsibility is caused more by a fear of failure.

SUGGESTED ACTION STEPS

1. One of the above scenarios for a new organization has occurred. Which of the scenarios in this chapter defines our situation most clearly?
2. Who are (or should be) the decision makers about roles and responsibilities for the organizational units involved in this change?
3. What are the primary topics that need to be discussed regarding responsibilities: products and services, customer–patient–client–student interactions, financial responsibilities, HR areas, etc.?
4. What are the next three steps for getting the areas of responsibility defined?

END NOTES

1. *USA Today,* June 15, 2005, MONEY section.
2. "Can Entrepreneurship be Taught," *MONEY Small Business Magazine,* March 2006, pp. 34ff.
3. Cummings, Thomas G. and Worley, Christopher G., *Organizational Development and Change,* 8th edition, South-Western College Publishing, 2005, 45ff.

Chapter 5

PERFORMANCE IMPROVEMENT AND GOAL SETTING: MAKING THE STRATEGIC VISION HAPPEN

INTRODUCTION

Chapter 4 concluded with a discussion of the importance of clarifying goals as a prelude to establishing responsibilities with "new" organizations. But clearly defined objectives or goals are often a need in organizations that already exist.

Making sure that the organization and its departments or other units have clear goals is an absolute requirement for defining performance, assessing current performance, and deciding what areas of performance should be improved. The most effective way for organizational units such as departments, teams, or new units to set their goals is for the larger organization (the company, corporation, or university) to have a fully developed strategic plan with "corporate" objectives. These organization-wide objectives form an effective context for the departments or units to decide on their own goals. A strategic planning process is discussed in Chapter 6.

While a clear and shared strategic vision from the leadership of the organization/team is the foundation for performance improvement, it is only the beginning. When the corporate vision defining products and services to be offered, markets served, methods of marketing, and the like are clarified, we have begun the trip toward performance improvement for the entire organization. However, many thorough strategic visions have stalled because people further down the organization did not understand the strategy, were unclear about their responsibilities, or did not buy into that strategy. Having goals and key performance indicators (KPIs) throughout the organization is essential in driving performance improvement toward achieving the strategic vision.

CHARACTERISTICS OF EFFECTIVE GOALS

Much has been written in general management literature about goals in organizations and teams. The following characteristics of effective goals are based on practical experience as well as a good deal of the literature. This discussion is presented in the context of performance improvement throughout the organization.

1. Goals should be tied to — or "cascaded from" — the detailed strategic vision developed and communicated by the leadership. There is more discussion about strategic planning and cascading goals in Chapter 6.
2. Goals should be *clear* and *specific*. (1) This means that it is important to avoid stopping with general statements such as, "Improve our sales" or "Make our communication better." General guidelines or intentions such as these can start a performance improvement effort but they are not themselves specific goals.
3. Goals need to be *measurable*, thereby providing the basis for evaluating performance at any point, including progress or lack of progress in performance improvement. Measurable goals, therefore, can help clarify the general guidelines listed in item 2 above.
4. Goals should be *achievable* — but challenging. (2) The common phrases capturing this characteristic are "stretch goals," and "out of reach but not out of sight." This characteristic is the basis for enhancing performance from where it would be without these stretch goals to a level of the best possible performance.
5. Goals should be *realistic*, meaning that they should not ask the team or individual with the goal responsibility to do the impossible, or to do something inconsistent with the nature and business of the organization. An example is to set a goal for an impossible schedule for rolling out a new product.
6. Goals should be *timed*, meaning they must be attached to deadlines for their accomplishment. Goals should also have action steps or, if they are major goals, project planning steps with dates for accomplishment.

The catch phrase used to communicate these principles is *smart goals*. Surprisingly, many organizations make sparse use of performance goals or objectives; and when they do, they are often not *smart*. This is true despite the fact that it is impossible to know with certainty the status of current performance unless goals and key performance indicators are set and used to measure performance. Sometimes organizations, even Fortune 500 companies, provide strategic goals or objectives from the top, but they are so general as to be difficult to define and measure. They therefore become less useful to teams and individuals below the top level in identifying what that team or individual should do to help the organization achieve its strategic vision and goals. This is a waste of resources and detracts severely from performance.

It is surprisingly common to hear leaders in organizations of all types say that formalized communication of the organization's goals is not needed because "everybody knows what we do here."

SOURCES OF RESISTANCE TO GOAL SETTING

We have found three primary reasons why organizations, teams, and individuals resist setting goals. The first, and perhaps most influential, is the fear of failure. The logic is that "if we are not pinned down to specific goals or objectives, we cannot be held accountable and punished if we do not perform." At the leadership level of the organization/team, the leader is sometimes afraid of knowing how things are going.

Experience has shown that when employees avoid goal setting, it is not usually motivated by laziness, but rather by lack of confidence or fear that leadership will use goals to punish or hassle employees. For people finding themselves in this situation, there is a better way to go than the avoidance of goals. They should work to develop specific goals for themselves or their team that meet the characteristics of *smart* goals. This makes it easier for the employee with a difficult boss to keep the conversation focused on specific performance areas. If their performance is meeting goals, it is much more difficult for unkind leadership to harass the high performer.

If the employee or team leader is not meeting the goals, at least the employee can figure out how to improve his own performance. Some organizational or team leaders or employees avoid setting goals so as not to feel bad about failing, but this can lead to the insecurity that comes from not facing our own fears. On the positive side, setting and reaching challenging goals can provide increased self-confidence and satisfaction. Otherwise, doubt and insecurity continues.

The second reason why organizational or team leaders do not use goal setting is lack of appreciation of the benefits from using such a system.

The benefits are numerous, and have an impact beginning from the top level of the organization to the functioning of teams and finally to individual performance. The list of benefits is provided below.

A third reason for resisting goal setting is the belief that in today's rapidly changing world, setting goals is impossible because of the need to be adaptive to changes. The reasoning goes something like this: "How can we set a target if the demands of the daily situation are difficult to predict? New business opportunities may occur, or customers may suddenly have new demands, or some technical breakdown may happen." The problem with this "go-with-the-flow" approach is that the organization becomes a slave to activity by focusing on reacting to changing circumstances. The best approach is to set goals at the organizational, departmental, and individual levels, but build into the goal setting plan methods and targets for being responsive to opportunities and crises. A case will help clarify this approach.

Scottrade Company

- Products and services:
  - Online discount trading in equities and other financial services
- Markets:
  - Provides equities trading opportunities throughout the United States and with selected international populations
- Structure:
  - CEO and eight top managers
  - Numerous managers divided into largely functional organization structure: IT, financial office, HR, etc.
  - Approaching 300 local branches at the writing of this chapter
- Primary issue and problems:
  - Wanted to have all departments with goals that helped the firm meet its strategically defined business interests

The large Operations department of this equities trading firm wanted to make sure that its programs met the business interests of the firm. The first step was to conduct its own tailored strategic planning process focusing on two primary areas of consideration: (1) the corporate strategic plan and (2) an assessment of its own departmental assets and challenges. In setting the corporate objectives, the department's leadership created a department process for dealing with external change. In this case, "external" also meant the firm as a whole because the firm was "outside" the operations department. This objective was as follows: "Establish goals in each department to keep Operations on track with Strategic goals" (Corporate). A process was followed for making this objective operational.

BENEFITS FROM ORGANIZATIONAL OR TEAM GOAL SETTING

Organizations, teams, and people tend to do things because they see desirable benefits. This is especially true of things such as initiating goal setting that require new behavior and attitudes. But goal setting has major benefits. "The research on goal setting... (findings)... are impressive in terms of the effect that goal specificity, challenge, and feedback have on performance." (3) The following are the primary benefits of a goals system, and these benefits seem to occur for organizations, teams, and individuals:

1. Goals provide focused action for achieving the organization's or team's strategy. Goals provide direction and focus for performance.
2. When teams and individuals have effective goals, they are clear about what is expected of them. So is everyone else, including his or her co-workers and manager. (4)
3. Goals are motivational, helping to elevate performance. This benefit occurs because some people like the feelings of success that come with reaching challenging and yet realistic goals. This has implications for hiring people who have high achievement needs that can be satisfied by setting and achieving goals.
4. Goal setting and the achievement of challenging goals build job satisfaction and self-confidence for those performing well. Obviously, it can also result in "downtimes" when goals are not achieved.
5. Setting goals can make work more "fun" and can help relieve boredom because the goals become like the score in a game, with the fun coming from meeting the challenge.
6. Goals provide the basis for measuring and evaluating current performance, and for evaluating progress or lack of progress with performance improvement efforts. This is a key point in our discussions of performance improvement at any and all levels of any type of organization.

CASCADING GOALS

As discussed previously, cascading goals from the top of the organization or team down to smaller units and individuals is helpful in driving performance. A case study will illustrate the point.

Organization: Tone's Brothers

- Products and services:
  - Processed spices

- Markets:
  - Retail chains
  - Food manufacturers
  - Cafeteria management companies
- Structure:
  - Plant of approximately 1000 people
  - About 200 managers, sales, and customer service personnel
  - The factory workers were unionized
- Primary issues:
  - Saw the need to connect corporate strategy to specific goals in production

This company wanted to reduce inventory, increase the number of inventory turns, and more closely connect product runs to the marketing strategy. To achieve this, we conducted a hands-on seminar with the production managers and supervisors using the following steps:

1. We discussed personality and behavior with a production team by providing each participant with a personality profile emphasizing their performance development. We used that information to discuss individual teamwork strengths and challenges. On their own, each participant identified things they would do to increase their team functioning as the group set its production goals.
2. We then reviewed the corporate strategic goals and systematically identified goals at the production level that would help them do their part in supporting corporate strategy.
3. We identified barriers or problems faced by this production team in achieving their department goals.
4. Finally, the team developed suggested action plans for overcoming those barriers.

This team of production department supervisors left the sessions with a list of specific goals and actions on which to work. As always, how well these were achieved would ultimately depend largely on department leadership support and persistence. In general, this particular company made a good effort at following through in achieving the production goals that were set in this process.

This case illustrates a useful process for cascading goals. First, leadership of the organization or team needs to make clear their strategic vision and resulting goals. Following that, each department or team within the larger unit should go through the process of identifying their goals and objectives and activities that are needed for them to play their part in the achievement of the overall strategy. When the department or team is a large one, and has a number of teams or other units within it, often there are three layers of objectives and goals: organizational, departmental, and teams or units within each department. The Scottrade case discussed previously in this chapter is a prime example of effective goal setting within a larger company.

SPECIAL CASE: LARGE TEAMS IN BIG ORGANIZATIONS

We have indicated that sometimes teams have the independence to consider developing their own strategic vision. When that occurs, they then provide a strategic vision to sub-teams within their larger team, and the cascading of goals process continues as usual. A brief case can make this more clear.

Organization: The Pro*Visions Pet Specialty Enterprises Sales Team for Ralston Purina, Co. (now a part of Nestle)

- Products and services:
  - Dog food
  - Cat food
  - Specialized pet foods, litter
- Markets:
  - Pet super stores
  - Regional chain stores
- Primary issues:
  - The need to clarify their goals and actions in support of Ralston's corporate strategy
  - The need to build the team toward increasing cohesion and performance

This team recognized that it had significant independence in making strategic decisions. Specifically, this team could decide which products offered by their company they wanted to sell to whom and which specific customers they should pursue. They were also responsible for deciding on the best sales and marketing strategy for their group and how to distribute the product once it was sold. While they had to develop this strategy consistent with the strategic direction and policies of the company, setting their own direction within that broader context was a key to higher performance.

The team began with strategic planning, developing a written plan involving a strategic vision of the desirable and intended future of their team. They then assessed their current situation and set immediate goals for achieving their selling strategy. What followed, however, was particularly thorough. This top team had additional meetings to discuss what each manager would do to carry his or her share of the strategic selling objectives. In addition, they identified and discussed what they and their direct reports needed to do to carry on their daily activities as part of the larger company. In short, each member of this top team committed himself and his direct reports to both strategic goals and daily operational goals.

CASCADING GOALS DOWN TO THE INDIVIDUAL

In the Ralston large team case discussed above, the planning process was completed when the team members developed and communicated individual action plans for achieving both the strategic and operational goals. Other individuals within their "sub-teams" also developed individual action plans and operational goals. Thus, the connection of the levels was complete — from corporate goals to large team goals, through the goals of the smaller teams, and finally to the individual level.

Many performance experts have recognized the influence of goal setting on team performance. (5) It is perhaps less well understood that goals, when developed correctly, also can have a positive impact on individual employee performance. Goal analysis is *sometimes* the technique used for developing performance for individuals, but in general, organizational or team leaders are much quicker to set goals for teams. Many performance management systems (appraisals) used for managing individuals in all types of organizations lack clear goal setting at the initial planning stage. Goal setting is absolutely essential for developing the individual employee when the employees and their manager plan work during the performance development cycle. Chapters 11 and 13 discuss this in detail.

Setting goals at the individual level is sometimes resisted because it requires that specific persons take responsibility for making things happen. At times, goals are effectively "cascaded" from the top level to the team or sub-team level, but are not integrated into individual action plans. Sometimes this is an oversight on the part of those responsible for moving the goals through the various levels of the organization; sometimes it comes from resistance by individual employees fearing accountability. However, the benefits of goal setting listed above are no less true for the individual employee than they are for the total organization or the teams that comprise it. Sometimes performance improvement requires taking a risk and committing to a performance goal even at the risk of not making that goal.

Resistance to taking responsibility for goals on the part of individuals may be caused by a desire to avoid accountability. But as often as not, it is caused by the goals simply being "forced" on employees rather than being discussed and agreed to in a collaborative manner. Leaders sometimes fear collaborative goal setting because they believe employees will "sandbag" by setting goals too low. Actually, our experience and numerous research projects have indicated the reverse. Employees are more apt to set goals too high, making them difficult to reach.

As the Ralston Purina case above illustrates, there are many different types or categories of goals. There should always be financial goals, even for not-for-profit organizations such as government agencies, because they still have budgets and related fiscal responsibilities. In fact, the most commonly found goals in all types of organizations are related to their budgets and the expense and income targets. However, other goals should exist as well. Some examples of other goals include production goals, selling goals, hiring goals, goals for creating new products and services, customer service goals, and human resources development goals.

MEASURING PERFORMANCE: GOALS AND KEY PERFORMANCE INDICATORS

Just as the strategic vision is the starting point for performance improvement and setting goals, the strategic and operational goals that flow from strategy are the starting point for measurement of performance. A goal is a desired end result. A major goal may have a set of shorter-term, or smaller, goals leading to the desired end result. The shorter-term goals still need to be *smart*. For example, if the end or major goal is developing and rolling out a new product by a certain date, then one of the shorter-term goals might be hiring a qualified research and development engineer.

There is always a series of important activities required to reach goals, short-term or longer-term. If the activities are not undertaken successfully, the goal will not be accomplished. Measurement of success in important activities leading to goals is done through key performance indicators (KPIs). KPIs are a limited number of performance measures that answer the following questions: are we doing enough of the right things to reach our goals? If not, what are the corrective actions to be taken? The primary goal for a call center might be, for example, a measurable level of customer satisfaction with the results from calls they have made to the call center. A related KPI could then be meeting a minimum level of "dropped" calls, those not answered before the customer gets tired of waiting and hangs up.

While measurement of performance at the organizational, team, and individual levels is necessary for performance evaluation and performance improvement, no organization or team need spend huge amounts of time or resources on collecting data for measurement of goals or KPIs. Only a few primary goals and "key" performance indicators need to be chosen for measurement. For example, if the goal is a certain level of sales volume, then the KPI may be new leads. If the goal is a certain level of employee morale measured by a survey, then the KPI may be undesired turnover. The principle here is Pareto's famous 80/20 rule. Simply stated, the famous principle is that 80 percent of what is important can be measured through sampling 20 percent of these important dimensions.

There are a number of key concerns when setting up measurement. One requirement is picking the right factors to measure and not measuring less important factors. Somewhat surprisingly, it is important to avoid measuring too much. Picking the right factor to measure is important for a number of reasons. First, it is important to measure factors where trends in the right direction on the KPI are indicative of progress on the goal they support. In addition, when organizations or teams use a KPI to measure activity, they are reinforcing it in the minds of the employees, and therefore they need to choose measures that are central to organizational success. It is also always true that organizations and teams get more of what they are measuring. To use the call center example again, if we measure how quickly our customer service reps answer the phone (e.g., less than three rings), we will get quick answering of the phones.

Establishing long-term and shorter-term goals permits measurement of performance through assessing goal achievement. Establishing KPIs, when we correctly identify the factors to be measured, also permits measurement of performance leading to goal achievement. Thinking back to the Model of Planned Change in Chapter 3, it should be clear that performance measurement is important, although in varying degrees, in all four stages of the model.

During the first stage of the GMPC, entering and contracting, organization and team decision makers are going to be clearest about the need to get involved in deciding on performance improvement if they have a number of significant measures of current performance. One measure most organizations pay attention to is their turnover rate for employees. Another set of measures organizations frequently keep is crucial ratios of some type. For example, an organization making heavy use of advertising will frequently measure the advertising cost per new customer or new student at the university.

During the second stage of the GMPC, discussions about how to proceed with diagnosis and what areas of the organization should be reviewed should focus early on existing performance measures. During diagnosis, new additional data may well need to be collected, but the organization's preexisting data on performance is a good place to start during these initial discussions.

Diagnosis is the most important and the most difficult step in deciding on current performance and deciding on the need for performance improvement actions. When we come to understand what areas of performance need review, and what models of effective performance apply, then the need for additional data may become apparent. There are four potential sources of data in any organization or team: (1) surveys, (2) observation of performance, (3) interviews, and (4) data already existing in the organization ("intrinsic data"). Decisions about which sources to use are a purely practical matter driven by decisions on areas of concern, what data already exists, costs, and existence of standards or models of effective performance.

A BRIEF CASE STUDY: ANOTHER MODEL OF EFFECTIVE PERFORMANCE

Organization: An Engineering Fabrication and Process Consulting Company

- Products and services:
  - Fabrication of process control instruments, analyzer systems
  - Consulting in mechanical, electrical, and chemical processes
- Markets:
  - Manufacturing companies
- Structure:
  - Approximately 170 employees
  - A small number of divisions: one focused on manufacturing, another on consulting
- Primary issues and problems:
  - Dramatic growth through doubling their size in the previous five years
  - Recently purchased by a huge company with resulting problems of adjusting strategy and procedures

After early conversations with this CEO/owner focusing on the above issues, we jointly reviewed the Organizational Success Model (see Chapter 2) to decide whether his primary issues were strategic or operational. We concluded that the issues were principally operational, with employee morale and relationships between offices and divisions being central. We then talked about the following Model of Effective Performance (6):

Effective Performance = Clear and Shared goals +
Good Attitudes + Knowledge/Skills + Motivation

The interviews with leadership provided data indicating that the primary issues were attitudinal. The specifics of the attitudes and morale held by the 170 or so employees, however, needed further clarification and definition before diagnosis could be completed and action decided upon. We developed a preliminary intervention that included collection of survey data from the employees regarding challenges, opportunities, and morale. The top team from all divisions and offices then discussed what to do about significant attitudinal problems.

As this brief case shows, data collection and measurement is often useful during planning and intervention, the third stage of the Model of Planned Change. During the intervention stage, the effort should be to directly tie measurement to performance improvement using the data to trigger improvement actions. So, one can see that in the above case, the engineering company's top team had the opportunity to use data on morale and attitudes in its efforts to improve these factors. The goal was to enhance communication and collaboration between divisions and groups of employees. In this case, negative morale and distraction caused by the takeover had impacted performance.

The final stage of the Model of Planned Change is evaluation and institutionalization of the desired changes. This, of course, only occurs if there is a performance improvement intervention. Clear goals and KPIs provide the basis for measurement, which helps us decide what in performance has been improved and, therefore, what we want to keep or "institutionalize."

Part III of this book discusses institutionalizing performance improvement; that is, making it an ongoing part of the organization. Because this chapter focuses on goals and measurement, however, it is important to make one major point very clear: *specific goals and related KPI measurements are essential for making performance improvement interventions last.* (7) A brief case will make this concept clearer.

Organization: Scottsdale Securities, Inc.

- Products and services:
  - Online discount broker providing equities trading
  - Broker-assisted trades as needed
- Structure:
  - Approximately 100 corporate personnel
  - Approximately 100 branches spread throughout the United States
- Primary issues and problems:
  - Lacked shared strategic vision
  - CEO, who was the founder and primary owner, had at least a partial vision of what he wanted to accomplish, but it was not fully developed and not completely understood or accepted by his corporate and branch managers

- Actions:
  - Conducted a thorough strategic planning process with the 20 or so top managers at the corporate level

The leadership of this organization agreed to engage in strategic planning. The early discussions with this leadership group indicated that the diagnosis of the need for strategic direction was correct. There was a lack of shared strategic vision among the leadership group. The CEO and his top advisors understood that agreement on a strategy, along with specific goals or objectives to accomplish that strategy, was essential to continued growth and success in the firm. The CEO could not do it by himself and he was smart enough to know it. Nineteen strategic objectives requiring immediate action were established at the end of the strategic planning process. Individual and team responsibility for these objectives was established primarily through the top managers volunteering to champion strategic goals or objectives they felt best able to accomplish.

Strategic goals or objectives are basically oriented toward long-term development of the organization. For this discount brokerage firm, daily operational issues, such as customer requests or technical problems for their automation-dependent business, often led to periodic delays in progress on their strategic objectives. However, the leadership of the organization continued to review and evaluate progress on each of the objectives. Within a year, all but one of the 19 strategic goals were either completed or had resulted in major progress toward achievement.

Approximately three years later, Scottsdale Company changed its name to Scottrade Company, largely for marketing reasons because their corporate headquarters was in St. Louis, Missouri. Each year the leadership of this company has reviewed and updated its strategy, setting new strategic objectives in the process. As of mid-year 2006, the company had tripled its number of branches from the date of the original strategic planning. It has become a large and very successful company.

THE ROLE OF FEEDBACK

When an organization establishes its organizational goals and then "cascades" those goals and key performance indicators, it is setting the basis for becoming increasingly successful, as has been true for Scottrade and other organizations discussed later. Performing all the work required to accomplish the organization's strategy sets up the possibility and requirement for feedback to those responsible for accomplishing the goals and supportive actions. Feedback to teams and individuals can be troublesome to those worried about not achieving their goals, but it is essential in evaluating current performance and identifying any need for performance improvement. In addition, those doing well at achieving the departmental, team, or individual goals need to be recognized and rewarded. A key principle of motivation, often ignored by leadership, is contained in the following phrase: "Rewarded behavior tends to be repeated."

Feedback to a team or department about how it is accomplishing its goals should be done in a way that allows all members of that team know how their group is doing. A quote from teamwork expert Montebello succinctly states the point (7):

> "Feedback has powerful motivational and developmental value for teams. It lets the team and the members know how they are doing so they know what to do more of, less of, or differently. Feedback can be about performance results, variances from plan, coordination with other units, customer satisfaction, or effective and ineffective behavior of the team. Depending on the type of feedback required and desired, there obviously are different sources such as internal customers, management, suppliers, and the team members themselves. But for feedback to be effective, it must be timely, specific, balanced and candid."

How much more effective will the feedback be when it is tied to specific team goals and KPIs? The same is true for performance at the individual level. Another performance expert notes (8):

> "One fact that researchers have demonstrated repeatedly over the years is that goal setting improves performance. ...Goals act as motivators, and research in both applied and experimental settings shows that difficult goals produce higher levels of performance than either no goals or simply instructions to 'do your best'. ... When workers know a particular level of performance is expected of them, they are motivated to try to reach that level, although it may be difficult. In addition, when group members are committed to a goal, performance of the group improves."

As the above quote from Montebello (7) indicates, how feedback is provided is directly connected to how it is accepted or resisted. Our suggested criteria for providing feedback to teams and individuals about their performance are as follows (9):

1. Feedback should be specific and, where possible, tied to specific goals or KPIs. Numerical data and specific examples are helpful.
2. Feedback should be sincere, reflecting honest attitudes and opinions from the provider of the feedback. Few things are as frustrating

for employees as having a manager who just read a new book or attended a management seminar and starts to provide ineffective feedback he or she does not really believe.

3. Making feedback timely is also critical. Receiving feedback on an occurrence some weeks past reduces the impact.
4. Feedback should offer suggestions for performance improvement where there are deficiencies. This is often a problem for managers providing feedback because they may not have an understanding of how performance can be improved. Obviously, referring the employee needing help with performance improvement to someone with experience in developing people is one possibility. In addition, sometimes setting specific goals for what performance improvement actions should be taken will lead both the manager and the person receiving feedback to find ways to make improvements.

Feedback is more apt to be taken seriously and serve as a trigger for action when the team or individual is committed to the goal and measured activities being discussed. Previously we argued for using a collaborative goal setting process to enhance commitment to the goals and the motivation for achieving them. While there is some evidence that employees can have as much commitment to assigned goals as to ones they help set, their participation in goal setting usually has no negative consequences and often helps with acceptance or motivation. For example, it is well established that commitment to a goal works best when the goal is consistent with the employees' values and when they believe success in achieving the goal is possible. Active collaboration between the employees and manager while setting goals allows for the opportunity to discover if this goal meets these requirements, while simply assigning the goal may not.

A brief case illustrating an effective and collaborative goal setting process followed by feedback illustrates a number of the points discussed in this chapter.

Organization: A Privately Owned Real Estate Company Managed by the Owner

- Products and services:
  - Brokered real estate for buyers and sellers
  - Purchased land and built apartments and other buildings for investment and sale
- Markets:
  - Homeowners selling or buying real estate
- Structure:
  - Multiple offices in small towns in a primarily rural area

  - Corporate staff including owner/manager and financial executive who managed budgets and conducted research on interest rates and local economic trends
- Primary issues and problems:
  - Owner/operator wanted to make sure that all branches of his geographically separated organization were motivated for achievement
  - Owner/operator also wanted to have a strong indication of probable financial performance for his total organization based on input from each office and real estate agent

This CEO/owner recognized the importance of gaining commitment and motivation to company goals. The process he used for achieving this was to meet with each individual real estate agent and their local office manager to set sales goals for each agent. The CEO had information developed by his financial vice-president pertaining to projections on interest rates, which dramatically affect real estate sales, and other occurrences in the local economy such as openings of new large businesses. The goal setting discussions were collaborative and, as often as not, the CEO and local office manager recommended reducing the sales goals suggested by the employee agent. They did not want to encourage frustration by letting these aggressive agents set goals that were both out of reach and out of sight.

The goals formed the basis for feedback to individual agents. The combination of goals for the agents in a specific location formed the basis for the goals for the local office, and the total of all the office goals combined to form corporate goals. This permitted the CEO to make financial decisions about buying and building real estate for his company investment.

It should be remembered that commitment to a goal is not the same thing as motivation to achieve it. But motivation to achieve a goal will usually not be strong if the goal is not accepted as desirable. So commitment is necessary for excellent performance, but commitment is not sufficient to result in high performance. What else is needed? In general, the employees must come to believe that working hard to achieve an organizational/team goal for which they have responsibility will lead to their receiving rewards and recognition they value. More about this in Chapter 9 and the discussion of leadership, teamwork, and their role in performance improvement.

CONCLUSION

Performance requires at least three elements: (1) smart goals that employees are committed to and are motivated to achieve; (2) key performance measures on progress in goals achievement; and (3) the resources to achieve the goals. These resources include tangible factors such as equipment and financial support, as well as effective leadership, which helps develop and clarify goals. However, it also includes having the right people in the right jobs. In Chapter 8 we discuss hiring, and throughout Part II we discuss helping people achieve the knowledge and skills needed for success in doing their work.

Hiring, goal setting, and training for job-related knowledge and skills are all essential to successful performance and improvement. However, these are "operational" actions, which best exist as direct expressions of the organization's strategic plan. We turn now to Part II and a discussion of the initial step in performance improvement — strategic planning.

SUGGESTED ACTION STEPS FOR ORGANIZATIONAL DECISION MAKERS

1. Discuss the following questions with your top advisors: Do we have clearly established strategic goals? If so, do we communicate them well to those further down in the organization?
2. Consider the following questions. What key performance indicators (KPIs) do we have? Which ones do we need? Do our KPIs measure the primary dimensions determining our successful performance as an organization? If any are missing, what are they? Are any of our KPIs not essential to our success?
3. Does our performance management system have goals and KPIs included?
4. Answer this question through input from employees: What strengths and problems exist in the ways our managers provide feedback?

END NOTES

1. Szilagyi Jr., Andrew D., *Management and Performance*, Scott Foresman and Company, 1981, p. 134ff.
2. Aldag, Ramon J. and Stearns, Timothy, *Management*, 2nd edition, College Division, South-Western Publishing, 1991, p. 428.
3. Robbins, Stephen P., *Organizational Behavior*, Prentice Hall, 1998, p. 180ff.
4. Aldag, Ramon J. and Stearns, Timothy, *Management*, 2nd edition, College Division, South-Western Publishing, 1991, p. 428.
5. For example, see Cummings, Thomas G. and Worley, Christopher G., *Organizational Development and Change*, South-Western College Publishing, 1997, p. 99.
6. This model is used by Achievement Associates, Inc., in many individual and team diagnostic situations.

7. Montebello, Anthony R., *Work Teams That Work*, Best Sellers Publishing, 1994, p. 322; and Aldag, Ramon J. and Stearns, Timothy, *Management*, College Division, South-Western Publishing, 1991, p. 428.
8. Cummings, Thomas G. and Worley, Christopher G., *Organizational Development and Change*, South-Western College Publishing, 1997.
9. Montebello, Anthony R., *Work Teams That Work*, Best Sellers Publishing, 1994, p. 322.

PART II

PERFORMANCE IMPROVEMENT: TAKING ACTION

Chapter 6

STRATEGIC PLAN FOR THE ORGANIZATION: WHERE IT ALL STARTS

CONNECTING STRATEGY TO PRIOR DISCUSSIONS

Part I focused primarily on an overview of performance for organizations, teams, and individuals. It provided a definition of performance, discussed why organizational leaders seek to improve performance, and provided a useful approach to performance improvement (the General Model of Planned Change, GMPC). Part I concluded with a discussion of goals and key performance indicators (KPIs) as essential elements for both measuring current performance and establishing performance improvement.

A brief summary of previous main points will help readers make the transition to this chapter:

1. Chapter 1 pointed out that many organizational or team leaders feel a powerful need to engage in performance improvement. On the downside, however, often the motivation of leaders to improve performance and the selection of programs used is triggered by mundane considerations such as the "latest and greatest" performance intervention. Part I used an introductory discussion of the role of the strategic vision of leadership in deciding the direction and nature of the organization or team, and therefore what type of performance is needed.
2. Chapter 2 took the concept of strategic vision and direction further, indicating that the ideal situation is one in which the

leadership develops a thorough and specific strategic concept. Following that, leaders must develop major objectives to accomplish their strategic concept and vision, and communicate those objectives to the rest of the organization. What occurs then is that levels of the organization/team below the leadership begin to understand the overall direction of the organization and can set goals, standards, and models of effective performance for their own areas. This permits the measurement of current performance and the ability to identify performance gaps between desired and actual performance.

Unfortunately, in the real world, getting leadership to clarify the strategic direction is frequently the most urgent need. As often as not, organizational leaders have not clearly and thoroughly laid out their strategic plans.

3. Chapter 3 discussed methods for deciding on performance improvement; whether it is needed; and if so, what interventions are best (GMPC). One major type of performance intervention, as demonstrated in the Organizational Success Model (Chapter 3), is strategic planning. Other performance improvement interventions, which focus on the operations of the organization, are most effective when they are developed, managed, and improved in light of the overall strategic plan.
4. Chapter 4 discussed performance improvements in different types of "new organizations," including those resulting from mergers and buyouts.
5. Chapter 5 discussed goal setting and key performance indicators, these goals and measures having "cascaded" from the strategic vision of the leadership.

This chapter focuses on a recommended process for creating a strategic plan, including the vision of the future, a mission statement, and strategic objectives for performance growth and improvement.

RECOMMENDED PROCESS FOR CREATING ORGANIZATIONAL OR TEAM STRATEGY

The strategy of an organization, or an independent team, is a *thorough and detailed statement of what the leadership desires and intends for the future of the organization*. It is the organization at its best, as defined by the leadership. The following is a description of what the strategic plan includes. The next four sub-sections of this chapter discuss a process for developing the strategic plan. (2)

Component 1: The Strategic Timeframe and Identification of the Driving Force

First, the strategic plan begins with two essential decisions: the strategic timeframe and identification of the organization's driving force.

The strategic concept statement, discussed below, is a thorough written statement of the desired and intended future of the organization. It is a statement about a future set of conditions for the organization. At least to some degree, the organization or team is not currently everything the leadership desires or intends. Therefore, the leadership group creating the strategic plan needs to pick a point in the future, usually three to six years ahead, to build its future strategic vision. As organizational leaders often say, "It takes time to get there." That time is the strategic timeframe.

There is no objectively right or wrong date for the strategic timeframe. Usually, organizations pick the date that marks the beginning of their fiscal year. Other than that, there are only general guidelines. A strategic timeframe of less than three years is too short because it becomes more like long-range planning or budgeting, two years-as-usual put back to back. More than six years is usually a problem for U.S. leaders because of their tendency to focus on short-term operational problems. Decision makers choosing a strategic timeframe often use considerations such as general length of life for product lines, churn or turnover in major markets, or how long the CEO has before retirement. Other considerations that often occur are rate of technological change and availability of capital.

One error in perspective that causes strategic planning teams difficulty is the difference between creating the strategic vision and "predicting" the organization's future. Creating the strategic vision is a proactive approach, deciding what the desired and intended future of the organization will be. It is a statement of faith in the leadership's ability to make the organization the best it can be.

This proactive creation of a detailed vision of "the best" organization is also not a utopian dream. An experienced team will inevitably use its sense of what "our business will be like" in creating of the vision, as well as the other steps in the strategic planning process. It is just that predicting what will happen is not the primary focus of strategic planning. Setting objectives and goals regarding what the leadership wants to happen is the primary focus.

The *driving force* is a much more complicated decision. It is also a decision that the leadership team needs to make careful choices. There are some key basic points about driving force that should guide the driving force choice:

- In large part, the essence of any work organization is the products and services it provides and the markets it serves. The remainder of the organization (technology, marketing approach and techniques, financial and human resources, structure and culture) supports the creation, marketing, and delivery of the products or services to the clients and customers. Thus, the most fundamental strategic decision is: what should the scope of our products or services and markets be?
- Determining the scope of the organization's products or services and markets affects the entire nature of the organization. If one makes a strategic decision to offer a new product line or drop a current one, one will be making decisions affecting human resources, marketing strategy, and production. If one decides to begin marketing products in a different country, or even a different state, at minimum one has made decisions impacting the distribution systems one uses and the marketing reach one seeks to accomplish.
- The driving force is the single strategic element that determines and drives the organization's total strategic concept. The driving force dictates future products and services; it is a major determinant of strategic decisions and future resource allocations. Driving force is the central theme and focus of the organization or team. The choice of driving force is a conscious decision made during strategic planning.
- Driving force is defined as that which determines the products and services and the markets of the organization. Ironically, while there are a number of driving forces possible, products and services offered and markets served are the most common organizational driving forces.

If an organization uses "products and services offered" as its driving force, it is saying that a well-defined array of product lines is the central focus of its organization. In this case, people in the organization will spend a great deal of energy and resources marketing those products and services everywhere deemed appropriate. New product development outside the basic product lines is de-emphasized, while product variation with the primary product lines does often occur. Growth of the organization comes from relatively stable lines of products and services, and marketing is emphasized by the organization taking its products and services into new markets.

In the 1980s, Anheuser-Busch Corporation (ABC) had a wide variety of products and services: theme parks, wine coolers, snack foods, luxury liners, a major league baseball team, and, of course, beer. At some point, the leadership made the strategic decision that the beer products were to again become the focus. In the terms used herein, ABC decided that its driving force was products: the many beers they were manufacturing. That led ABC to sell off many existing businesses outside its primary product line, which had always been beer. Now, however, beer was clearly the "king" of its products and the focus of its strategy.

A few additional points about the ABC case are important here. People drink beer at baseball games, in theme parks, and on luxury liners. But beer is a "secondary product" in those events, not the primary product. When Anheuser-Busch began eliminating peripheral businesses, it also expanded its marketing of the primary product line, beer. When the company refocused on its beer products, they were selling that product in a handful of countries. As of 2006, the company sells extensively in Mexico, Asia, Latin America, and much of Europe. ABC has also added to its primary product lines. However, except for theme parks, its products are primarily beer.

If an organization uses "markets served" as its driving force, it is committed to understanding its defined market as well as possible, and then doing what it can within its capability and capacity to provide products and services needed by that market. In this case, the organization spends an unusual amount of time conducting market research and developing new products to satisfy that market. This is what is sometimes called using a "market niche." In this driving force, the market drives the product line, and research and development of products is the focus. In the "products and services offered" driving force, the products determine which markets are the focus. Examples of "markets served" organizations include Gillette Company and Merrill Lynch & Co. Inc. In both cases, their approach is to define and understand their market's demographics and provide a multitude of services to meet that market's needs. Gillette, for example, offers a wide variety of different products: razor blades, deodorant, shaving cream, after-shave, razor blades, and the like. The technology for making these products is very different; some are liquid and some are metal. What makes them similar is their end users — "the market served."

Other options for driving force exist, although they are less common. "Technology" is the driving force for Texas Instruments and Microsoft. A highly structured "distribution system" is the driving force for McDonald's Corporation. And a distinctive "selling method" is the driving force for Avon, Amway, and Spiegel Inc. "Capacity" is one of the most unusual driving forces. Moore Business Forms is an example of a "capacity" driving force, in that this company is large enough to be a major supplier to the Internal Revenue Service.

- Deciding about driving force is probably the most complicated and most important strategic decision for an organization. Many organizations/teams have never thought about making a driving force decision, and thus they float between multiple driving forces, "trying to be everything to everyone." The problem with that approach is lack of focus on core business, diluting of resources, and confusion about what opportunities to pursue and which ones are not worth pursuing. An organization, no matter how large or extensive its resources, cannot afford multiple driving forces within the same organization.

- Changing driving forces can be done consciously but requires great consideration because of the costs. If, for example, Anheuser-Busch were to decide to focus on "markets served" instead of "products and services offered," it would need to define its primary markets clearly. If it defined that market in a specific way, which its large and talented marketing department could do, it might focus on "beer drinkers in the Americas and other developed nations." That definition of a primary market and the "markets served" driving force would suggest that Anheuser-Busch should begin conducting or confirming research on non-beer products and services desired by that market. The brewery is famous for having excellent marketing management and research capabilities, so it could undoubtedly accomplish the research task. The strategic problem would be to develop the capacity and capability to produce, market, and distribute new products and services for its market. One can envision a major shift toward a number of products: beer tappers and portable pre-fab bars for home and business use, information on food and beer combinations, T-shirts and beer mugs, and perhaps even a "hangover" medicine. The point is that these are very different products from the primary focus on beer and would require new production capability and production support services.

A brief actual case study helps illustrate the importance of the driving force decision. This case study was used to make a different point in Chapter 5.

Organization: Tone's Brothers

- Products and services:
  - Processed spices
- Markets:
  - Retail chains
  - Food manufacturers
  - Cafeteria management companies
- Structure:
  - Plant of approximately 1000 people
  - About 200 managers, sales, and customer service personnel
  - The factory workers are unionized

- Primary issues:
  - Organization had hired a new Human Resources director and significantly added to the HR staff

The new Human Resources director, well educated and experienced in organizational performance, realized that he and his team members needed a planned direction. They understood the process for strategic planning described in this chapter and decided to use it for planning the direction of the HR department. Their CEO had given them a great deal of freedom in charting the future direction of the new HR department.

In the beginning of their strategic planning process, they chose their strategic timeframe, which was four years into the future. Then they discussed what their driving force would be. First, they considered products and services. They thought they could identify the products and services they should provide during the strategic timeframe. They reasoned that following their listing of the desired and needed products and services, they could begin to plan all the actions they needed to have the capacity and capability to deliver those products and services.

They started to face a crucial issue as they continued to discuss the driving force. All of them, with the exception of one HR staff member, were new to the company. Even the new HR director was new to both the company and the industry (spice processing and marketing). In addition, they were probably going to be providing HR services for plants or company locations other than where their plant was located. All of this made the Strategic Planning team feel like they needed more information about the company and its needs before they could choose HR products and services.

They began to discuss a second driving force choice: markets served. They realized that they would have to work hard to understand what was needed for each internal customer group.

The team chose the second approach. Their reasoning was that "We know who needs Human Resources services better than we know what they need, except within standard categories of services such as benefits and some type of training. We build the capacity to develop whatever is truly needed, and we can develop the capability to provide what is needed. We do not have a substantial list of existing services, and have a great opportunity to build products and services in response to our internal customers' needs (markets)."

The strategic planning team then identified their internal markets, primarily managerial groups and frontline workers. Following that, they began to create their Strategic Concept Statement, the second component of the strategic plan following driving force decisions. This particular team used good judgment in choosing the driving force. They had a sense that the decision would affect much of what they did for a long time. They also realized that the right choice of driving force would provide a major opportunity for building their new department.

Remember that this was a department within a larger corporation. In the context discussed in Chapter 4, it was a "new organization." But because it was a brand new department with new employees and management, they could choose their own direction. Most importantly, the company CEO wanted them to chart their own direction. In this situation, a department has the ability to do some strategic planning regarding its own direction.

Before discussing the next element of a strategic plan, the Strategic Concept Statement, a word of caution about the "markets served" driving force is useful. Markets, clients, citizens, or customers served (the label depends on the organization doing the planning) is not the same as customer service. Good customer service is about things such as reasonable credit and product return policies, on-time delivery, repair services, and friendly and responsive interactions with sales and service people. "Markets served" as a driving force uses the needs of the market to determine what products and services one will create and provide. This is a different decision from those about how customers are treated.

Component 2: The Strategic Concept Statement

The second component of the strategic plan is the Strategic Concept Statement.

The Strategic Concept Statement is a detailed word picture that describes the desired and intended future of the organization or team. It is the written description or "vision" of the unit as leadership would like it to exist. The leadership team creating the strategic plan makes the following decisions:

> "By year 20… (i.e., three to six years into the future), what do we desire and intend to have as our primary products and services and our secondary products and services? What products and services do we choose to evaluate during the timeframe? By year 20…, what do we desire and intend to have as our primary markets served and our secondary markets served?"

The discussion surrounding these questions produces a picture or vision of what the leadership wants to provide and to what markets. The same questions and resulting decisions are also applied to five other strategic elements: (1) sales method (or marketing in some organizations, communication in government agencies, etc.); (2) distribution system; (3) technology; (4) major sources; and (5) capacity and capability.

The result from discussion of the strategic vision (concept statement) is a detailed description of the desired or intended organization. It is much more than two paragraphs of "who we are," sometimes called a mission statement or values statement. It is a plan for the future, a model of "effective performance" for the organization. In addition to narrative descriptions of desired and intended products and services, technology and the like, it contains key performance indicators (KPIs) for the entire organization or team that measure the progress toward the strategic vision. Numerical values for each of the KPIs are established for each year during the strategic timeframe, and are changed as experience dictates.

In various places throughout previous pages in this book, we discussed organizations and independent teams as if they were alike strategically in terms of setting direction and goals for team members. If a team has the independence to make choices about its own products and services, the markets it serves, and its technology and major sources, then it can develop its own strategic plan. Examples of this situation existed in the Tone's Human Resources case. The main requirement for the strategy of a team within a larger organization is that it must also account for the strategy, desires, and wishes of the larger organization of which it is a part. That is, however, not a great deal different from a total organization being concerned about the strategy, desires, and wishes of its major customers.

All organizations have a strategic concept or vision, meaning a more or less clear definition of products and services, markets or clients being served, etc. But as discussed previously, that strategic vision is often vague, contained only in the mind of the top leader, different for each of the top leaders, and communicated poorly. The process of strategic planning is aimed at *creating a shared picture of the desired and intended future organization or team.* It is also wise for leaders of organizations to communicate this shared picture or vision to all of their employees.

Component 3: Evaluation of the Current Situation

The third component of the Strategic Plan is the evaluation of the current situation, both internally and externally.

This is where internal strengths (S) and weaknesses (W), and external opportunities (O) and threats (T), are listed and discussed (i.e., the SWOT analysis). The best approach is to evaluate the items of the current situation regarding their importance in achieving the strategic concept statement (vision). That is why it is necessary to create the vision of the future before evaluating the current situation. A simple three-factor rating scale is used (1 - 2 - 3), with 3 indicating major significance of the situational item and 1 indicating minor significance.

Major elements of the internal situation, which are reviewed in light of the strategic concept statement, include facilities and equipment; systems, methods, and procedures; leadership, management, communication, and control; human resources; current products and services; sales and marketing; and financial systems and status. Each area is taken in turn, and suggestions for the significant strengths or weaknesses to include come from any of the individuals on the strategic planning team. The entire team uses the scale above for rating the strategic importance of the item. The data for identifying items within the financial situation of the organization usually comes from the leadership of the organization.

The choice of who should comprise the strategic planning team is a crucial one. The best guideline is to select people who will be particularly significant for making the strategy a reality. Most often this will mean selecting people in the top leadership and management groups. But sometimes there are people within the organization who have special knowledge or experience in an area crucial to organizational success. As such, the leadership team can choose to involve them in the creation of the strategic plan.

There is another opportunity for broadening the strategic involvement of people not in higher leadership levels. The opportunity is to generate input from employees throughout the organization, getting them to focus primarily on what they know best — operational issues. As part of the SWOT analysis, managers can either interview or survey employees on operational items they see as strengths or weaknesses in the areas where they work. These items are then evaluated as to their importance by the strategic planning team. As discussed elsewhere, employees usually know more about the daily operations of the organization within their work area than do their managers. It is the wise manager who makes use of that knowledge, strategically as well as on a daily operational basis.

In most cases, employees other than leadership will be especially insightful about strengths and weaknesses internal to the organization. People with customer contact may also see external opportunities and threats in products and services, sales method, or technology. Gathering information from others in the organization on both strategic opportunity or threats and operational topics broadens interest in the strategic plan among employees. This broadened interest usually establishes a good basis for communicating the completed strategy to the entire employee group.

The evaluation of the strategic importance of the SWOT items suggested by other employees is done by the strategic leadership team because strategy is the responsibility of organizational leadership, no matter the type of organization. Some leaders wonder why, because strategy is a leadership responsibility, they should communicate that strategy to the entire employee group. The reason is simple. The strategy is best achieved when everyone in the organization is doing their part to achieve it — leadership, production, marketing and sales, customer service persons, and those in finance. To accomplish their part of the strategy, employees need to know at least the strategic vision and the strategic objectives.

Evaluating current organizational strengths and weaknesses in terms of their importance in achieving the strategic objectives has a number of benefits. First, any organization has many items that are seen as assets or liabilities on a daily basis. One of the interesting side benefits of this process is that it drives identification of these SWOT elements (assets and liabilities) that all organizations live with. Some things are irritations or the basis of some satisfaction. Other aspects of the organization are strategically important. Evaluating these strengths and weaknesses clarifies which is which.

Component 4: Strategic Objectives

The fourth component of the strategic plan is strategic objectives (macro goals).

The process for defining these goals is as follows. The strategic concept statement (vision) is where the leadership group desires and intends the organization to be at some point in the future. The current situation is where the organization or team currently is in a number of areas. The items, strengths, and weaknesses within these areas have been evaluated in terms of their strategic importance.

The strategic objectives are the immediate goals and actions established to make progress from where the organization is currently to where leadership desires and intends it to be in the future. The objectives are the actions to achieve the strategy. The objectives are the centerpiece of organizational performance for the organization as a whole. This is true regardless of the type of organization: for profit or not-for-profit, manufacturing or service, university or business corporation, whatever industry they are part of.

Strategic objectives are set for immediate action, with people identified as responsible for achieving the objectives. Preliminary action steps for each objective should be created by the strategic planning team. They have established the strategic objectives based on a shared vision of the future that they have created. The initial action steps can usually be deduced from the strategy, and they flow quickly from the leadership team. But the initial steps for each objective are just a start.

The key is for each strategic objective to have a champion, someone accountable for making the objective happen. Of course, that champion may need a number of people to help in achieving the objective. The selection of other people to work on each of the objectives is also a key to success in making the objectives happen.

Figure 6.1 provides a graphic description of the Strategic Planning Process discussed above.

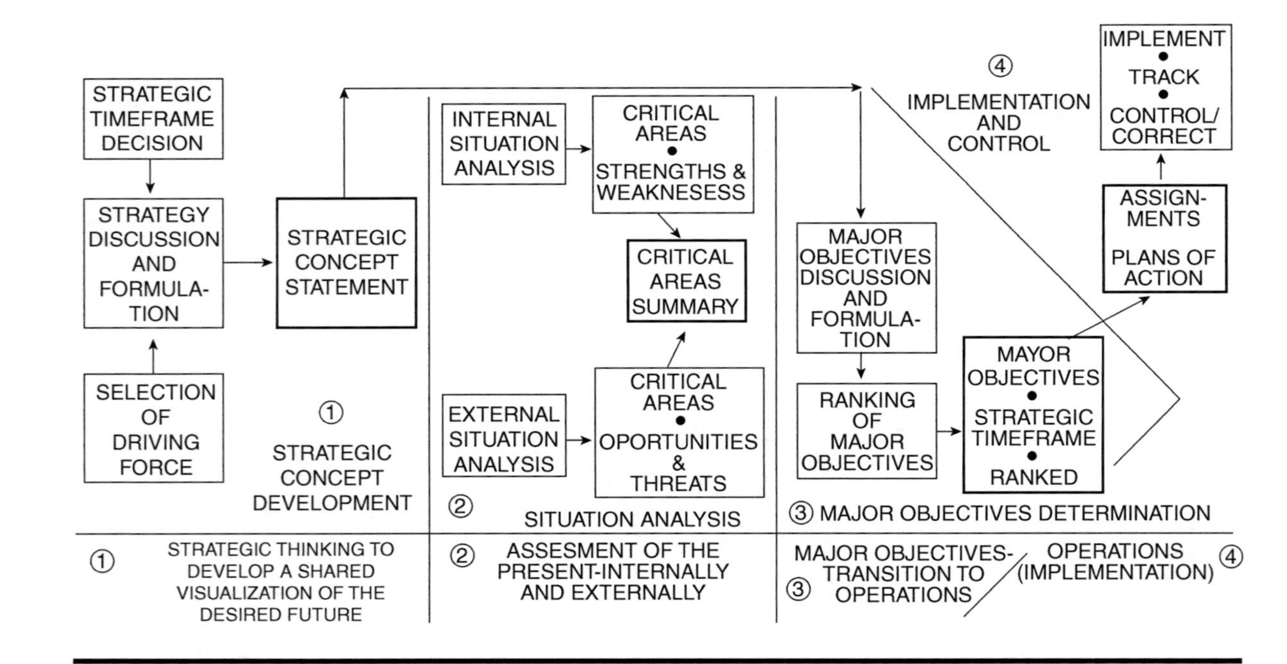

Figure 6.1 Strategic planning process. (Copyright 1994. Achievement Strategies Division, Achievement Associates Inc.)

CASCADING GOALS REVISITED AND EXTENDED

At numerous places earlier in this book we discussed the importance of using leadership's strategic vision to cascade goals throughout all levels of the organization. After the strategy is developed, cascading should start with a presentation to the entire employee population of the basics of the newly developed strategic plan. In general, communicating the primary elements of the Strategic Concept Statement (vision) and the Strategic Objectives is the most important. The details of the SWOT evaluation usually provide more information than is needed by the average employee. Items in the SWOT analysis that are significant to a particular department within an organization should be discussed with the leaders and employees of that department even if other SWOT strengths or weaknesses are not included in discussions with a particular team. For example, if the SWOT analysis determined that current equipment used for communication is deficient, then technical and finance departments should be informed. If the SWOT analysis revealed customer service as a major strength for the organization, discussions with the call center or other customer service personnel stressing continued emphasis on this area would be essential.

Experience shows that the single most important reason why organizations or teams fail to make a strategy happen *does not come from weaknesses in the plan itself.* Weaknesses in the plan can be overcome during the regular meetings of the Strategic Objectives team. As a part of their meetings, the team will conduct a review of progress on strategic objectives and make thoughtful adjustments in the plan as necessary.

The single most important cause of strategic failure is a lack of persistent efforts at achieving the strategic objectives. This failure is typically due to a deficiency in leadership. When a strategic plan fails because the objectives are not accomplished, the failure is due to a lack of leadership emphasis on getting the objectives accomplished. When the objectives are achieved and are successful in moving the organization in the right direction, it is due to both the emphasis on and support of the strategy by leadership and the diligence and hard work by those throughout the organization actively involved in the strategic objectives.

There are a number of ways to manage actions aimed at achieving strategic objectives. The most common is for the CEO of the organization to manage the actions through his or her top team. Two brief case studies will clarify this. The first is a case discussed originally in Chapter 5.

Organization: Scottsdale Securities, Inc. (now Scottrade Inc.)

- Products and services:
  - Online discount broker
  - Broker-assisted trades as needed

- Structure:
  - Approximately 100 central corporate personnel
  - Approximately 100 branches spread across the United States
- Primary issues and problems:
  - Lacked a shared strategic vision
- Action:
  - Conducted a thorough strategic planning process with approximately 20 managers at the corporate level

The CEO had a large number of direct reports. In the first strategic planning process, the CEO and his corporate department leaders engaged in strategic planning that resulted in 19 strategic objectives. These objectives covered a number of strategic areas; the creation of major new products and services; the development of new technology as a foundation for online brokers; and development of human resources, creating a training function focused on improving customer service in the branches. The CEO and his HR manager provided oversight of the actions on these objectives by having persons responsible for each objective develop and communicate action plans and steps for that objective. They then conducted meetings with the strategic objective leaders to discuss progress on each objective.

While the meetings were not as frequent as the CEO and his top advisors wanted, the process of managing the achievement of these objectives from the top down worked. Within approximately one year of the creation of the strategic plan, all but about five of the nineteen major Strategic Objectives had been completed, or there had been significant progress on them. The strategic plan has continued to be a major part of the growth and success of this firm close to a decade after the original strategic planning process.

In today's fast-paced world, it is inevitable that pressures of daily operations interfere with the long-term performance improvements that come from a strategic plan. This is even truer when the leadership of the organization is more complicated than at Scottrade, Inc. (a single CEO who is also the major owner). The following case study illustrates challenges resulting from multiple ownership. This brief case study also demonstrates how one can partially overcome issues deriving from a diverse leadership by having a strategic plan.

Organization: Medium-Sized Law Firm (Roberts, Perryman, Bomkamp & Meives, P.C.)

- Products and services:
  - Litigation: personal injury for defense
  - Business legal services

- Markets:
  - Casualty and property insurance carriers
  - Small to medium-sized businesses
  - Any self-insured organization within its geographic market
- Structure:
  - Headed by six partners, with a managing partner who was also a major owner
  - Approximately ten additional attorneys
  - Approximately twenty other staff members, from paralegals to clerical support
- Major issues and problems:
  - The managing partner believed the group lacked a common direction and had major disagreements over marketing their services, finances, profit sharing, and hiring decisions

Enough other partners agreed with the managing partner's perspective on the organizational issues that they decided to engage in strategic planning involving the six partners. After a lengthy process and many discussions, the partners agreed to five major objectives, including a marketing and sales plan, major technology improvements, financial procedures and systems development, and a legal skills development program. Two of the five strategic objectives were accomplished quite quickly; three were slowed by disagreements between the partners. Three of the partners were persistent in moving the three unmet objectives. This persistence helped the firm achieve significant progress on the final three objectives within 18 months of the creation of the strategic plan. After their original strategic planning process, the law firm updated their strategic plan and added new objectives.

Some organizations manage actions for achieving the strategic objectives by cascading those objectives to teams and making the team leaders responsible for their accomplishment. In this case, having the top leader highly involved is somewhat less critical. This approach works best when the organization or team is made up of distinct groups with clearly defined functions. Again, a brief case study will clarify this.

Organization: Development Department of an Average-Sized Private University (Webster University)

- Products and services:
  - Graduate education
  - Undergraduate education
- Structure:
  - Organized into three teams by function and client group

- Primary issues and problems:
  - Needed a strategic plan with objectives that could help ensure that the teams were working together and in the same strategic direction
- Action:
  - Developed a strategic plan with 14 strategic objectives

The strategic leadership group developed a strategic plan and then made each of the three top team leaders individually responsible for moving the objectives ahead. By the end of the first year after development of the strategy, more than half of the strategic objectives had been completed. While the top manager was involved in providing advice and assistance to the team leaders, the team leaders were each principally responsible for completing those objectives with support from people within their own team.

How active the top leader needs to be in working and achieving the strategic objectives will vary in different organizations. Still, the support of those working on these objectives from the CEO or president is crucial to success. Part of the success of organizations comes from cascading the strategic objectives from the top manager to the appropriate department or team leaders. The final step in this cascading of the strategic objectives is to put some part of the strategic objectives in the action plans of the appropriate individuals.

Individualizing the organization's strategy helps make performance improvement permanent. This step also increases the probability that the strategic objectives for each year will be accomplished. The ideal is for the team or department leader to take responsibility for making sure the objectives are accomplished. However, managers and employees throughout the organization should have the primary responsibility for working and completing the strategic objectives.

The performance action plan for designated individuals should include appropriate goals for achieving strategic objectives. *In short, all the individuals' strategic action plans and goals taken together comprise the strategic goals and action plans of their team. And all of the various department or team strategic action plans comprise those of the total organization.*

MAJOR PROCESS ISSUES

Any major effort at performance improvement, of which strategic planning is an example, requires dedication and persistence. Two dynamics can occur that will jeopardize the effectiveness of strategic planning:

1. Hard work and persistence are essential for both the creation of an effective strategic plan and achievement of the plan's objectives. The effort at developing the Strategic Concept Statement requires vision and creativity as well as open communication. Evaluating the current situation requires persistence, determination, and continued openness in communication. Establishing the strategic objectives and planning the actions for those objectives requires endurance.
2. Deciding who should be involved either in the planning team or in working a strategic objective can be complicated and is a crucial decision. Two principles are important here. First, anyone should be included who will be important in making the strategy happen. Second, any exclusions should be done on a rational basis, not a "political" one. Being "popular" or "unpopular" is not a major factor in one's importance strategically.

CONCLUSION

Developing and communicating the strategic direction of the organization or team is the first obligation of leadership. The second is making sure it happens. Communicating the completed strategy to the entire group is a start for the organization on clarifying its current performance and knowing what improvement is needed. The remainder of Part II provides performance improvement interventions to help organizations, departments or teams, and individuals perform successfully. This includes successful performance in strategic actions as well as daily operational action.

SUGGESTED ACTION STEPS FOR ORGANIZATIONAL OR TEAM LEADERS

1. Make sure you understand what a strategy is and what it is not. Distinguish strategy from yearly operational plans, statements about value and missions, and short-term goals.
2. If your organization has a strategy, answer the following question: Is our strategy understood throughout the organization?
3. If your strategic direction is understood and accepted throughout the organization, make sure you have a process for regular meetings with those responsible for achieving the strategic objectives. Recognize that there are daily operational issues required of all leaders, but the strategic objectives should also be emphasized and have deadlines.

4. If there is insufficient agreement on strategy (e.g., products, services, markets served, marketing and selling system, etc.), then get your leadership group together to go through the development of a strategic plan.
5. Make it happen.

END NOTES

1. Meister, Jeanne C., *Corporate Universities: Revised and Updated Edition,* ASTD, McGraw-Hill, 1998, p. 93ff.
2. Many of the concepts for the strategic planning methodology discussed in this chapter were influenced by a class text in the field. Our thanks to Tregoe, Benjamin B. and Zimmerman, John W., *Top Management Strategy,* a Touchstone book, Simon & Schuster, 1980.

Chapter 7

BUILDING A LEARNING ORGANIZATION: IMPORTANCE AND METHODS

REVIEW AND LINKAGE

Part I of this book discussed a number of key points that are recapped here:

- There is a powerful need for performance improvement in the United States. This need is being triggered by a number of social and economic factors that have been in existence for a number of decades: intense and growing competition, the changing nature of work and the U.S. workforce, and the importance of building shareholder value in corporations, among other factors.
- Decision makers from all types of organizations or teams initiate performance improvement efforts with motivations that will often influence the success of the efforts at improving performance.
- There is a process we recommend for deciding on current performance and whether to engage in performance improvement. Following this process increases the chance of improving performance.
- Performance starts with the strategic vision of the leaders of the organization or team. That vision includes the products and services of the organization, their markets or client groups, and other major parts of the organization that assist it in delivering what the organization provides to those who value it. Many times, this strategy is poorly detailed or not shared with others in the group or organization who are needed to make it happen. Clarifying and detailing a thorough strategic vision is achieved through open discussions with those knowing the organization or how to improve performance.

- Once the leaders' strategic vision and the accompanying strategic goals are communicated to the organization or team, the opportunity exists for evaluating current performance and identifying performance requirements at the operational level. That is done through goals, standards, key performance indicators (KPIs), and models of effective performance. These operational measures provide the basis for knowing the current status of performance.
- The process we recommend for evaluating and developing performance (General Model of Planned Change, GMPC) assists organizational leadership in its efforts at diagnosis through intervention and evaluation and institutionalization of improvements. The diagnosis stage of the process seeks to identify areas of performance gaps between desired goals and standards and actual achievement.
- Performance requirements are most effective when they are cascaded as goals or standards from the strategic vision of the leaders. Efforts at performance improvement that are underway also need the same leadership support and specific goals as do other measures to evaluate progress and make it permanent.

When used correctly, all of these key points made in previous chapters allow the organizational or team decision maker to answer the following questions:

1. What is good performance in my group or organization?
2. In what areas is our performance good enough to be acceptable, at least for now?
3. In what areas do we have performance deficiencies or gaps?

The eminent Peter Drucker recently wrote a brief passage that captures the essence of Part I (1):

> "Every organization operates on a Theory of Business, that is, a set of assumptions as to what its business is, what its objectives are, how it defines results, who its customers are, what the customers value and pay for. Strategy converts this Theory of the Business into performance."

However, strategy and achievement of the strategic vision is achieved by people throughout the organization or team performing at work in ways that accomplish goals, meet standards, and are consistent with models of effective performance in areas of importance. Thus, people must both know how to do their work and want to do so. Because strategy is dynamic, and there is constant and accelerating change in the environments of organizations, people must keep up with the performance requirements in their organizations and teams. *Ultimately, the need for performance improvement by all employees is continual.* At the organizational level, that is what is meant by "lifelong learning."

THE LEARNING ORGANIZATION: A STARTING POINT FOR PERFORMANCE IMPROVEMENT

Many hard-driving organizational or team leaders (often motivated by success and money, politically astute, and somewhat cynical) might begin this chapter with a good deal of skepticism. "Learning organizations" sounds academic or at least extravagant. What does this really have to do with performance improvement anyway?

A vast number of experts in management, performance improvement, and related fields have been arguing for at least the past decade that we live in an information age. Among other things, this means that what we know is a primary key to how well we as individuals, teams, or organizations succeed. As Peter Senge notes, "The organization that will truly excel in the future will be the organization that discovers how to tap people's commitment and capacity to learn at all levels of the organization." Senge uses a quote from the head of planning at Royal Dutch Shell to dramatize the importance of learning: "The ability to learn faster than your competitors may be the only sustainable competitive advantage." (2)

Chapter 1 began with a discussion of the powerful motivation many leaders feel for performance improvement in their organizations. This need, felt by many organizational leaders, has led to efforts at improving organizational, team, and individual performance in all kinds of organizations. As a result of the vast amount of work in performance improvement efforts, we have acquired extensive experience with a number of interventions available for improving performance: strategy, restructuring, performance management systems, and training and development, for example.

These interventions all require learning on the part of those involved in the interventions. In addition, the rapid change in technology, the way employees do their work, and the requirement of being responsive to clients or customers means that employees at all levels have to continuously learn. This is a requirement for organizational life today, whether or not the organization is involved in formal performance improvement programs. Employees at all levels must continuously learn as part of doing their jobs well. This is at least part of what Drucker and others mean by "knowledge workers."

DEFINING A LEARNING ORGANIZATION

The following summarizes some of the main points that come both from research and experience in building learning organizations.

1. Workers at all levels and in all kinds of organizations have to know a great deal about a wide array of areas: how to operate computers, the recommended methods for dealing with customers, the most recent improvements in technology for performing the jobs they do, and what is expected from them in working with others as a team member or manager (supervisor). In general, what employees at all levels need to know grows and changes constantly.
2. The "lean" organizations so common today often reduce the influence of supervision and management for workers at all levels, meaning that workers at all levels are expected to be more "self-directive" and "figure it out" for themselves. This increases the need for those workers to be even more productive than before, doing the work of more than one person and learning what they need to know to do the job.
3. Lean organizations also means that many workers must develop self-management knowledge and skills.
4. Workers at all levels must be prepared to change jobs, add significantly to the jobs they do, and even change careers. Learning is a lifelong process.
5. The role of managers and supervisors is undergoing change. Managers, teams leads, and supervisors are required even more than before to mentor or teach those under their direction in order to enhance performance. Increasingly, workers solve problems, and managers and supervisors help them learn how to problem-solve. This often includes finding the resources for the problem solving.
6. Organizations that are leaders in building learning cultures usually insist that their employees have knowledge and skills in both the core competencies related to their current jobs and also in other areas. Organizations often also want their employees to understand the organization's culture, values, and traditions. As well, employees are required to have knowledge regarding competitors and the best practices at least within the areas of expertise required of the employee. (3)
7. Some organizations have come to recognize that employee learning and skill development in "personal" areas is important because our personal life usually affects how we perform at work. While corporations are careful not to legally violate the privacy of their employees, "employee assistance programs" and training in personal

areas is occurring in all types of organizations. For example, education on personal finances is common, and courses on goal setting and time management are everywhere and can be applied to personal areas of life as well as to work.

8. Organizations benefit by providing many learning opportunities for their "knowledge workers," partly to improve productivity and partly to keep them from going to work elsewhere.
9. Learning activities should be connected to a clear understanding of what the organization needs to accomplish (its strategy). In addition, learning activities should be closely connected to what each individual employee should know how to do, needs to know but does not, and needs to know now and in the near future.
10. The points made above are important to work organizations of all types, not just large, for-profit corporations. Learning the basic knowledge and skills to do the job has always been important in job success. In today's complex, highly competitive, and rapidly changing world, continuous learning is absolutely essential to individual, team, and corporate performance.

To summarize the above, a high-performing organization must have employees, essentially from the top to the bottom of the structure, who have the following characteristics. First, they recognize the need to continuously learn and develop their knowledge and skills. Second, valuable employees know they will be asked to add to what they do as part of their role in the organization. They may have to take over someone else's job temporarily or as a permanent move. And perhaps most importantly, they will often be asked to teach others about some element of job performance.

Knowledge employees are willing and motivated to go beyond the core requirements of performing their current jobs. They are usually willing to engage in activities ranging from coaching and counseling to helping design the strategy of the organization/team. The organization must hire people and develop people who have these characteristics. It is then essential that the organization provide the learning-oriented employee with the learning opportunities. Those employees who have the need and motivation to learn and do new and additional work will not be satisfied if they are hired and promised an opportunity for learning, variety, and growth, and then those opportunities are not made adequately available. When these growth opportunities are made available, the organization will thrive. Everyone will be working at an energetic and productive level. A case study from the rapidly changing industry of banking and financial services will illustrate the reality of a productive organization.

Organization: Mid-Sized Credit Union

- Products and services:
  - Checking and savings accounts
  - Individual and family loans, cars, houses, etc.
- Markets:
  - Four branch locations in a city of approximately 2.5 million people
- Structure:
  - A community-based credit union with citizens in the service communities comprising the Board of Directors
  - Small management team with strong CEO and branch managers
- Primary issues:
  - Credit Union had suffered business loss from two forces. One of their primary branches was in a rapidly declining community that was suffering an out-flight of its primary employers and their middle-income employees. The second occurrence that harmed them was the deregulation of the financial services industry, meaning that banks, credit unions, and even brokerage firms were all increasingly providing much the same services. Therefore, smaller organizations such as this credit union often had trouble competing.

The credit union had a number of options. It could merge or become an acquisition of a larger financial institution. The second option was to disband or await bankruptcy and be forced to close. The third option was to dramatically change how and where it was doing business.

The leadership of the credit union chose the third option. In summary, they engaged in the following actions:

1. They developed a strategy that involved expansion into additional geographic areas populated by citizens with the socioeconomic profile of people attracted to this credit union.
2. They dramatically expanded their Human Resources function with specific attention given to hiring new employees with a willingness to learn new skills and attitudes necessary for success in the increasingly competitive financial services industry.
3. They developed a comprehensive program for culture change within their organization, with increased focus on customer service and appropriate selling of multiple services to customers (cross selling).

THE MODEL FOR GROWTH: LEARNING AND SKILL DEVELOPMENT

In performance diagnosis and improvement, models play a critical role in helping us obtain some perspectives on a number of crucial questions.

First, what kind of performance is occurring now? Second, what is causing the current performance level — both those factors helping our performance and those limiting that performance? Third, where and how can we improve performance?

A simple distinction often used in management consulting is the notion that performance occurs at three levels: (1) the organizational level, (2) the team or department level, and (3) the individual level. While this distinction can be useful in defining answers to the above three performance questions, they also tend to ignore a basic reality. That reality is that in the final sense, *all performance comes back to the individual level.*

As is discussed in Chapter 10, teams, particularly permanent ones like departments or ongoing policy teams, have an impact on individual performance and behavior. But teams are still comprised primarily of individuals, and the individuals comprising them impact what teams do and how well they perform. When team membership is changed, teams usually change significantly in many aspects of what they do and do not get done and how they function.

The same is true for organizations. Organizations often outlast their original members and founders. Their goals, structure, norms, rewards, and sanctions impact the behavior of their members. But at any one point in time, the individuals within that organization and its departments impact the organization and its various units. This is part of why hiring (discussed in Chapter 8) and team membership are such crucial decisions. When a production team in a manufacturing company changes its maintenance staff, for example, the entire process will be impacted as maintenance routines are changed to accommodate the new personnel. The same types of changes occur when people within any department are changed, or when existing individuals change what they are doing. Therefore, it is essential that we have a model for understanding how people learn and grow, and why sometimes, unhappily, they choose not to learn and grow.

The Model for Growth described below and in Figure 7.1 helps with understanding what areas of the individual impact performance and the related need to learn. This model is particularly useful because it provides us with an understanding of both those things in individuals that can support learning and growth and those aspects of the individual that can work against learning and growth. Most importantly, this model, developed by one of my associates, provides an understanding of self-development techniques that impact learning as well as other areas of individual behavior. In summary, the Model for Growth helps us be specific about individual learning and skill development, and methods that can be used for that learning and skill development. It also can alert us to styles of learning and barriers to learning.

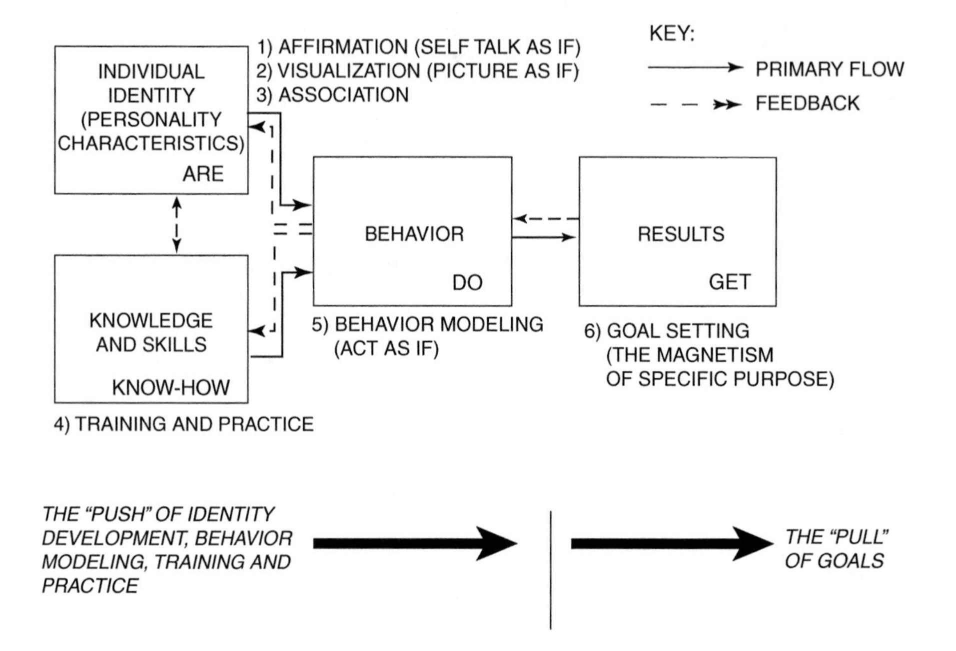

Figure 7.1 Model for Growth.

The Model for Growth describes the individual human being as follows:

1. The first element of the individual person is *identity.* The identity of an individual includes a number of aspects. One aspect of identity is personality, which is reflected in our behavior. Another aspect of identity is motivation, which includes values or what we think is important. Attitudes, how we see and evaluate our world, is also an aspect of identity. In summary, our identity is "who we are."
2. The second element of the individual is *knowledge and skills.* Knowledge is what we know — our theories, ideas, and concepts. Skills are the things we are able to do, frequently as a result of our knowledge about what we are doing. Knowledge and skills taken together are "know-how." To know how to operate spreadsheets on a computer, one needs some knowledge and skills in operating the computer, spreadsheets, and probably basic math. Sometimes people know something in terms of concepts but do not have the skills, or perhaps the motivation, to carry out that knowledge successfully. Many managers "know the theory" but for any number of reasons do not succeed in putting that knowledge into their management style.

3. The third element of the individual, according to the Model for Growth, is *behavior*. Behavior is what we do. Behavior is a result of who we are (identity) and what we know how to do (knowledge and skills).
4. The fourth element of the individual described in this model is the *results* individuals get from their behavior. Results for the person is a source of his satisfaction or frustration with his life.

In summary, who we are (identity) and what we know how to do (knowledge and skills) result in our behavior (what we do), which leads to our results (what we get).

The "results" referred to here include how we learn, what we learn, and what we do not learn. Subsequent chapters in this book discuss how the model for growth helps with hiring effectively and with helping people learn how to improve their performance as managers, team members, and sales and service providers. The remainder of this chapter discusses how we can know about an individual's identity, current knowledge and skills, style of behavior, and results in ways that help us know how to help them learn.

A number of learning programs and activities focus at least partially on the identity of the learner. Andragogy, theories about how to best teach adults, focuses on the significance of identity as a factor in learning. The principle used here is that different adults learn differently. Most importantly, we have fundamental tools that help us know how to maximize the individual learner's motivation and success at learning.

This is a brief sample of some of the better-known assessment tools that are related to a person's identity and their willingness and ability to learn.

1. The first element of the individual, identity, directly impacts what we learn. There are numerous personality assessments aimed at hiring or individual development that provide insight into a person's learning orientation and style (among many other individual facets). Examples include The Achiever, the Myers Briggs assessment, and the various DISC assessments. (See Chapter 8 on hiring and Chapter 13 on improving performance at the individual level.)
2. There are a great number of feedback instruments that evaluate a person's motivation, values such as honesty that impact giving and receiving feedback, and attitudes toward co-workers or the organization. Most morale surveys rely heavily on assessing the attitudes of respondents as an expression of their identity.
3. Many feedback instruments on leadership or supervisory style evaluate management philosophy and attitudes toward workers, such as theory X versus Y, and authoritarian, collaborative, or *laissez faire* styles of leading. Willingness to learn from others is encouraged or discouraged by our attitude toward them.

As discussed previously in this chapter, the second element of the Model for Growth is knowledge and skills, that is, what one knows (concepts and information) and what one can do (skills). Three key points related to this element of the individual's learning are important here:

1. If one understands something but cannot translate that knowledge into how one behaves, then it cannot help one's performance.
2. Conversely, if one can do something, but does not really understand it, then one will have trouble transferring that skill to different applications or situations. For example, if one succeeds at dealing with a difficult customer but does not understand what worked, then one will not have much success in transferring that successful performance to other customer situations.
3. The following are different things to be learned. Data is information, a factual representation of reality, specific details, and sometimes numerical representations of something that exists. Concepts are theories or ideas that organize data or representations about something, someone, or some behavior. Skills are the abilities to do something with concepts or data.

Many, perhaps most, learning programs and activities provided by organizations focus on the employees' knowledge and skills. Training and development programs for managers, customer service providers, and sales people, for example, teach concepts and techniques for performing in those roles. When teaching employees new computer systems, or conducting on-the-job mentoring and training, the goal is building knowledge and skills so that people can perform their jobs at a high level.

Translating knowledge, concepts, ideas, and information into skilled and effective *behavior* is one of the great challenges in training and development or any form of learning. Very often, things learned in the classroom are poorly transferred to the work situation, if at all. More about the specifics of that issue appears in Chapter 15. Here, the key point is that people learn to do things differently and more productively (skills) through practicing the most effective behavior. As the Model for Growth reminds us, our results come largely from our behavior. And our behavior comes largely from our identity and knowledge and skills.

Human identity is formed partly through conditioning whereby we develop attitudes, habits, values, and motivations through dealing with our environment. Developing new skills is, in large part, working to eliminate habits that do not work and replacing them with more desirable and productive habits. This is the methodological basis for the wise phrase that "Repetition is the mother of learning." Repetition over a period of time helps people learn and change behavior if they choose to make that change. Many organizations today have come to recognize that they benefit by helping their employees, at all levels, learn what they need to learn to be able to perform at their highest level.

Three cases from actual organizations having some success in helping their people learn in ways related to performance will illustrate some of the points made heretofore in this chapter. The first two cases were in the first edition of this book and have been updated. The third case is new to this edition.

BUILDING LEARNING ORGANIZATIONS: THREE CASE STUDIES

The three case studies that follow help illustrate how organizations can implement many of the concepts, tools, and techniques identified in this chapter. As always in this book, these are actual cases in which those of us at Achievement Associates have had direct involvement.

Case Study 1: Landshire Sandwiches, Inc.

- Products and services:
  - Packaged sandwiches, burritos, etc.
- Markets:
  - Convenience stores and gas stations
  - Miscellaneous quick food retailers, bars, etc.
  - Provide product to outlets in 16 states
  - In the past five years, the company has added selling directly to vending companies as an additional market
- Structure:
  - Centralized preparation and packaging of their product
  - Corporate headquarters located near the primary plant
  - Company has added a second plant in recent years for production of breads used in their sandwiches
  - Depots located strategically throughout their 16-state geographic market for short-term storage of products
  - Route drivers and supervisors deliver the product to retail outlets that sell to the consumers
- Primary issues and problems:
  - The owner and the CEO together had two primary elements in their visions of the organization: (a) they wanted to encourage the development and growth of their employees, and (b) they wanted to maintain organizational growth and financial success in a high-stress route business

The owner left operations largely to the CEO, the new president who handled the vending company aspect of the business. Other top managers were involved in the financial, technical, and sales operations of the business. The owner had strong feelings about treating employees well, as did the CEO and the president. They also were very successful as a business, adding outlets and even entire states in an ongoing manner. The addition of the vending business in the past few years had added dramatically to their growth in business. However, they knew that route businesses have characteristics that can produce intense problems. Their production and delivery schedule was unforgiving; every day their outlets demanded deliveries. Failure to deliver could result in the convenience store chain shifting to other food suppliers. The vending machine aspect of the business was slightly less pressure-packed than the convenience store aspect of the business.

In addition to schedule challenges, route businesses — particularly ones like Landshire with a large geographic area to cover — face problems that come from the physical separation of their employees. There were route drivers operating in numerous locations in each of these states, and local supervisors attempting to carry out corporate policy. Coordination, communication, and organizational clarity were special problems. The schedule and the geographic spread of Landshire's customers resulted in production, selling, delivery, and service needed at hundreds of locations across 16 states on a six-day-a-week basis. There was little time to spare.

The leadership of the organization knew that it had some performance problems coming from this geographic separation. These issues existed primarily in the attitudes and motivation of the route sales personnel, particularly those some distance from headquarters. Interpersonal conflicts in the organization sometimes took an unreasonable amount of time and attention away from production and delivery. Turnover, always high in route businesses because of the stress, was higher than necessary. Other operational issues also existed, despite the fact that great strategic positioning within a high volume growth industry (convenience stores) made them financially successful. The leadership decided to undertake a series of performance improvement activities:

- They conducted a small number of thorough morale surveys to define the specifics of their operational issues as seen by people within the organization. They wanted to learn more about where they were as an organization, so that performance improvement could be targeted and progress could be measured.
- One of the most dramatic results of the surveys was that there were leadership style issues — some lack of clarity about directions and policy, and the existence of pockets of fear of punishment if

things did not go well. Continuing these conditions was unacceptable to the three top leaders.

- As a result, Landshire engaged in a series of learning and performance improvement interventions aimed at improving leadership style throughout the organization and providing clearer direction whenever possible. These included:
  - The development of educational programs to help managers at all levels learn effective management and leadership styles, tools, and techniques. This included providing a number of on-site management development programs, as well as providing off-site educational opportunities.
  - Evaluating whether employees were using the management and leadership knowledge and skills that were provided. This has been done fairly regularly during performance feedback sessions from managers to their direct reports. Often, the direct reports are themselves managers of others. This adds a dimension to the learning requirements of those who have to do their own work and oversee the work of others.
  - Using a collaborative approach to establishing sales, servicing, and financial goals within each of the geographic areas comprising their total market. Feedback on achievement of these goals within each area is then regularly provided, and actions taken as needed. Recognition and reward is freely given when deserved. In summary, the managers and supervisors of the company have learned the techniques of goal setting and have become very skilled at it.

The executives at Landshire would be somewhat reticent to think of themselves as a "learning organization" in many of the ways listed previously in this chapter. However, they are headed in the right direction for a number of reasons:

- The top leadership of Landshire is committed to using learning programs and developing knowledge and skills as a primary way of creating the kind of organization they envision.
- They recognize that learning in many areas is essential for a successful organization, including managing, selling, conducting technical operations, and many more areas. They also realize that the learning opportunities they provide must be directly related to their strategy and the nature of their business. They recognize and have dealt with the need to provide learning opportunities to employees spread across 16 states, performing a wide range of functions.

- They have sought to tailor learning to the strategy of the business and to the knowledge, skills, identity, issues, and strengths specific to each team or individual involved in the learning experience.
- They have shown incredible persistence in providing learning opportunities to their employees, some eight years after their initial surveys identifying morale issues. In addition, they have been determined to reinforce learning through setting goals and providing feedback on how well people use what has been learned.

As of this writing in 2006, Landshire continues as a successful business. It continues to work on training and developing its people — in seminars, classrooms, and on the job. It has the most essential characteristic of a "learning organization." It knows that learning is a lifelong process, and it persists in finding ways to provide its people with opportunities for learning related to organizational, team, and individual performance and continuous efforts to improve performance.

Case Study 2: Anheuser-Busch, Inc.

Landshire Inc. has accepted the challenge of developing the performance of its widely dispersed people through learning opportunities. By comparison to the next organization, Landshire has relatively few resources to expend on learning. Anheuser-Busch, Inc. (ABI) is at the other end of the resource continuum, spending millions of dollars a year to provide learning programs and activities.

- Products and services:
  - Many lines of beer
- Markets:
  - Sells primarily to more than 1000 distributors throughout the world
  - Increasingly involved in marketing beer throughout the world: Mexico, Latin America, Asia, and Europe (as of the writing of this edition, ABI is considering beer sales in India)
- Structure:
  - 5000 employees producing, marketing, and delivering product
  - Plants across the country, with headquarters in St. Louis
  - Field sales force located in regions across the United States
- Primary issues and problems:
  - As of the mid-1990s, ABI had developed a group of training and development programs and activities that spread throughout the organization. Each of these efforts was largely developed independently of the others. The objective was to pull these units

together into a "corporate learning function" to accomplish the following goals:

- Make sure programs and activities were tied directly to corporate strategy
- Increase the reach and penetration of the collection of programs through coordination and sharing of efforts
- Increase the "relevancy," meaning tie the programs and activities directly to position related "core competencies" and other supportive learning opportunities

The decision makers at ABI created an organization reflecting a new learning alliance among some preexisting programs. The learning alliance, called The Anheuser-Busch University (ABU), had the following units:

- *College of Production:* engineering for the breweries, packaging units, corporate and plant engineering, and all technology teams
- *Busch Learning Center (BLC):* developmental experiences for sales and marketing, brand management, media groups, field sales, and distributors
- *Leadership:* leadership and professional development

The ABU has a number of features that make it a model for learning programs in larger organizations, including:

- The method of organization served the overall strategy of the corporation, which was based on a major emphasis on production quality followed by extensive sales and marketing. The leadership and professional area supported the emphasis on production and marketing, and that group had a significant, but smaller, role of the three units in the learning alliance.
- The learning activities conducted were based largely on an assessment of what individuals needed to enhance their long-term performance. The leader of that organization and the Busch Learning Center (BLC) constructed a map of the performance strengths of the units in which employees worked. This "strategic map" focused on the group's strategic plan, its structure, teamwork and team building, leadership strengths and issues, and review of processes. In addition, each individual who was to be served by the BLC conducted a self-assessment on required knowledge and skills for his or her position. Part of that assessment included how the individual stood in mastering that information and those abilities. Individual performance reviews, conducted with their managers, provided some input to their assessment of where they were

performing well and where they needed help. All of this resulted in the Individual Development Plans.

- The BLC, in particular, provided learning opportunities for important organizations in its distribution chain, which starts with the wholesalers and retailers receiving their products. This characteristic of providing learning to groups outside the employee population is a distinguishing feature of "corporate quality universities." (4)

Recently, Anheuser-Busch consolidated all training for both the beer company (ABI) and the overall corporation (ABC) into one coordinated program. This program, headed by Randy Roberts, is called the Beer Opportunities Training Center (BOT). The current training (2007) still makes use of the Individual Development Plans detailed in the above paragraphs.

Case Study 3: Scottrade Inc. (see Chapters 3 and 6 for data on Scottrade)

The primary learning-oriented challenge that this company faced originated in the kind of business it is in and therefore the knowledge and skills of people it hires. Basically, Scottrade depends very heavily on two groups of employees. First, it depends heavily on employees who know the stock market and are licensed, or can be quickly licensed, in the trading of stocks, bonds, and related services. The second group of employees critical to this company are people with knowledge and skills in information technology (IT).

This company has grown rapidly, especially in the past ten years. In 2000, it had approximately 100 branches and the same number of corporate personnel. As of 2006, there are close to 300 branches. In 1995, the average number of employees in each branch was two or three people. Today, that average is four to six employees, and there are 100 corporate employees in IT alone. This has meant that Scottrade has needed to help many of its highly technical personnel to learn how to effectively engage in "soft skill" areas: participation in annual strategic planning, managing others, functioning as supervisors or team leads, conducting performance goal setting and reviews, and working in teams.

The company originally approached the provision of learning opportunities to their people almost exclusively through the use of outside trainers and educators. Gradually, Scottrade built its internal capacity to provide learning for its people — both technical and "soft skills" learning. As of 2006, this company can boast of a high-quality "Scottrade University" with a broad-based program rapidly approaching what a "corporate university" should provide. Most importantly, the company works to meet the two requirements for effective leading within organizations. First, the company links its learning programs to the achievement of its strategy. Second, the company is working to tailor its learning opportunities to the individuals participating in its learning programs.

SUMMARY AND CONCLUSION

The three cases used here are about organizations with very different resources. That was done to make a point. If the leaders of organizations or teams believe that learning and developing their people is crucial to performance success, they can find ways to do it. ABI University has hundreds of people involved in conducting its programs. It has an educational satellite, a system established at the cost of millions of dollars. And it makes extensive use of computer-based training and interactive television.

Landshire conducts on-the-job training for production and sales and marketing employees, and finds external programs and resources for much of its learning effort.

Scottrade is somewhere between the two above cases regarding the resources it can and does commit to helping its people learn. It has a small but growing training department. It provides learning opportunities in many different subject areas, and it has multiple learning delivery systems, including its intranet. The three organizations, dramatically different in size, share a common characteristic: they all recognize the value of learning in performance improvement, and they all make a determined effort to maximize learning for the benefit of their organizations and their employees.

SUGGESTED ACTION STEPS FOR ORGANIZATIONAL OR TEAM LEADERS

Answer these questions:

1. What do we know about ourselves as people? What are our attitudes, motivations, and basic personality characteristics? How can we come to better understand who we are and how to get more out of our identity?
2. In what areas of work do we have sufficient knowledge and skills? In what areas do we need improvement in knowledge or skills?
3. What areas of behavior are most significant to our organization/team: managing, marketing or selling, customer service, technical support, financial, etc.? In which of these areas do we have sufficient positive models to move our behavior in the correct directions? In which of these areas do we need to find "models of effective performance?" How do we find them and then install them in our organization or team?

4. Are our goals clearly established at the organizational or team level and at the individual level? How do we increase goal clarity? Do we need to re-do our performance management system to increase the role of goals and KPIs?

END NOTES

1. Drucker, Peter F., *Management Challenges for the 21st Century,* Harper Business, 1999, p. 43.
2. Senge, Peter M., *The Fifth Discipline,* Doubleday Currency, 1990, p. 4.
3. See Meister, Jeanne C., *Corporate Quality Universities,* Richard D. Irwin, Inc., 1994.
4. Meister, Jeanne C., *Corporate Quality Universities,* Richard D. Irwin, Inc., 1994, p. 33.

Chapter 8

SELECTION, STRATEGY, AND PERFORMANCE: PROCESS, TOOLS, AND TECHNIQUES

INTRODUCTION AND OVERVIEW

Most organizational or team decisions makers understand, at least intuitively, that the selection of people for positions is closely related to organizational performance. After all, the best chance for excellence in performance is to have the right people in the right job. At the very least, it is clear to most organizational leaders that performance will suffer if they have the wrong people. (1)

However, the following issues sometimes cloud this intuitive wisdom about the importance of good hiring:

1. Many organizational leaders involved in hiring have not spent enough time and effort in understanding what their strategic plan and goals, or those of the hiring department or team, indicate about what type of person(s) should be hired. The linkage between what we are striving to be as an organization and whom we need in our organization is often not thought through.
2. Organizations often have trouble knowing what personality characteristics, behavioral traits, or competencies are required to do well in a specific position, as well as within the department or the organization as a whole.

3. Even when they are clear about the above, which is infrequent, those involved in hiring for organizations often have trouble matching the applicants' characteristics to what is known about organizational or departmental needs and job requirements.
4. When a good applicant is found, the match with job requirements is made, a hire occurs, and everyone is happy. Often, the happiness lasts for only a short period of time. As a CEO business associate once was heard to say: "I hired him and his ugly brother showed up." "Ugly" referred to performance once the person was on the job.
5. Despite having experience with and performance information about current employees, many organizations do an ineffective job in transferring and promoting existing employees. These motivational tendencies include placing a problem employee somewhere else in hopes he or she will do less damage or "straighten out." In the reverse, often organizations take people performing well in operations and automatically make them managers because they have operational skills (e.g., sales people becoming sales managers, a production person becoming a supervisor). In this case, many organizations depend *solely* on a person's performance record in one job although he or she is being considered for a very different position. Performing well in one position does not mean performing well in another position — even when it is in the same organization.

Using an effective selection process is critical in helping an organization avoid hiring mistakes like those listed above. When an effective selection process is used, organizations can dramatically provide the Human Resource basis for top performance. An effective hiring process should meet the following selection requirements:

1. The objective of selection or hiring is to find a person who "fits" *all* of the "identity" requirements of the position. This means the person should have the motivation, personality, attitudes, and values that make performing well on the job possible. If any of the identity characteristics held by the new hire are inappropriate for the position, performance will suffer. Toward the end of this chapter we discuss "normative profiling," an approach based on profiling identity elements that are related to job performance and then developing a process for measuring those elements in job applicants.
2. Managers are usually looking for applicants with the knowledge and skills necessary for quick ability to perform in the position. Some organizations believe that the identity factors listed above are the most critical considerations in hiring, and that when they

find the right hire they can teach the knowledge and skills necessary to the position. That approach is most appropriate when hiring for less complex or entry-level jobs. In general, the organization will want to hire people who already have knowledge, skills, and requisite experience when filling positions such as marketing or sales managers, top sales people, top managers, and technical or production employees. But while knowledge and skills are more important with higher-level jobs, the identity factors are just as important. Many times, a new hire moving into a top-level job within the organization struggles with the job and eventually leaves — not because of technical deficiencies, but because of a "poor fit with the culture."

3. Organizations and teams consist of interdependent workers. How well individuals perform at least partly depends on how they interact with and support their co-workers. (2) For example, the supervisor of the first stage of the production line may need to collaborate with the supervisor of the second stage to solve some processing problem. Hiring needs to pay attention to how the applicants will fit with the culture and personalities of the people with whom they will work. It is wise to involve selected co-workers in the evaluation of final applicants for a position with which they have major interaction.
4. Hiring is very expensive. Experience and research show that out-of-pocket costs for hiring a manager average about one year's annual salary. And if hiring is expensive, poor hiring is even more expensive. No one is quite sure how to precisely measure the costs of a poor hire when it leads to the person leaving later, but the calculation of costs should include rehiring expenses plus lost production during downtime.

While making sure that the applicant "fits" the job is important for the organization/team, it is also critical to the applicant. How many people have had their careers derailed and spent months, years, or decades unhappy at work because they were in the wrong position? One author notes that studies show that most workers would not choose the same line of work again. (3) All in all, it makes sense for those hiring and those being considered for a position to work together to decide if the "fit" is good for both.

SOURCES OF INFORMATION FOR HIRING DECISIONS

Hiring is more like gambling than it is like a science. This is true for at least the following reasons:

- It is sometimes difficult to know exactly what is needed for excellence on the specific job open for hiring.
- People are complex and often difficult to assess as applicants.
- Any new employee is influenced in his or her job performance by his or her personal life, which is difficult to assess during hiring and is unpredictable in any case. Things such as divorce, illness, and financial difficulties, to name just a few "personal problems," impact performance. The key is to improve the odds of making the right decision in the hiring gamble, which is principally achieved by *gathering as much information as possible about the applicant.*

The following are sources of information important in hiring decisions.

Step 1

The place to start is in gathering information and defining the "identity" characteristics (personality, motivation, values, and attitudes) required for success in the job. People who manage the position being filled can be one source, as can co-workers or others who have succeeded in similar positions. The next step is to identify the knowledge and skills, behavioral style, and output goals required for excellent performance in the position being filled. This is more than using the traditional job descriptions, which are often outdated because jobs change quickly. What is required is a fresh look at the job being filled. Developing a thorough understanding of the individual characteristics and competencies required by the job takes a systematic approach and a willingness to revisit the topic occasionally. (4)

The Model for Growth (Figure 8.1), discussed in Chapter 7, is useful for purposes of clarifying job requirements categories.

Information on job requirements for each of the boxes in the Model for Growth is important. Can anyone doubt that identity factors such as personality, motivation, and basic values play a part in performance? While the importance of knowledge and skills will vary with the complexity of the position, skills at dealing with people and working in teams are almost always important. In addition, a profile of behavior or style required for success in the position is important. Finally, some positions (production, project management, and selling for example) require a strong commitment to output goals and deadlines. There are methods available for gathering information on all of these areas of job requirements.

Step 2

Once the requirements of the job have been clarified, information about the applicant can be used to see how they fit with those requirements.

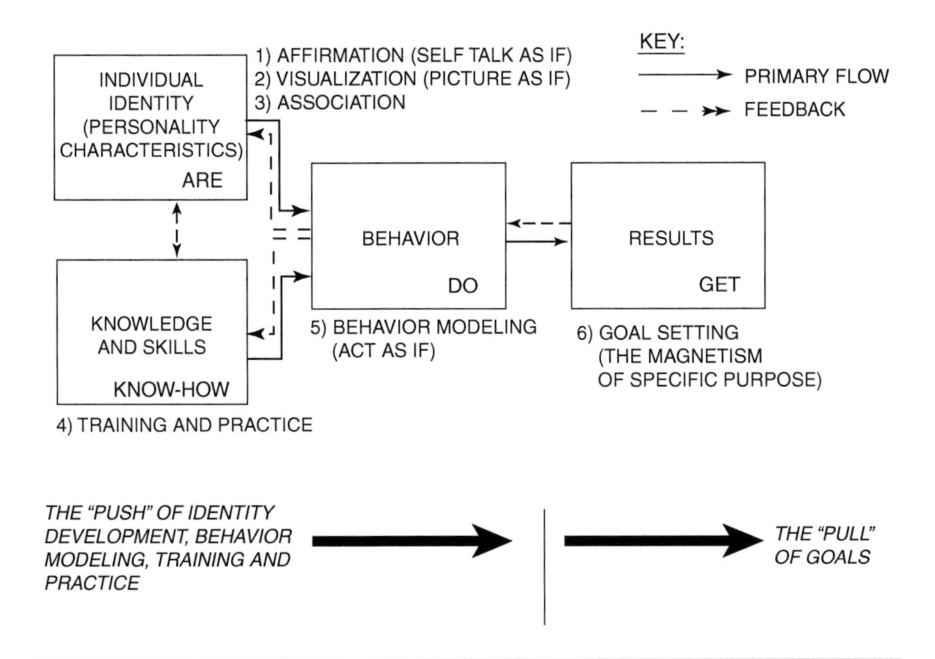

Figure 8.1 Model for Growth.

The source of applicant information always comes from some type of application or individual resume. Frequently, today's applications are submitted online. The information contained in the application and resume is used to identify applicants for the next step. The identification of these applicants is done by comparing the information they have provided with the job requirements defined in Step 1 above.

Step 3

When the review of the information on resumes and application forms has resulted in a reasonable list of potential hires, the next step is *checking references.* There are many challenges with reference checking, as most HR people know. Those challenges include the fear of lawsuits many HR managers or CEOs have, which leads them to be cautious about providing information. The basic recommendation is to use questions that are largely factual and to focus on the knowledge, skills, and goals of the applicant. So, in asking for references from a former employer, the following questions should provide some information. Does Mr. X have a good understanding of marketing research techniques? Does Ms. X usually meet deadlines in submitting reports? The one factual question that can be asked about overall performance that is frequently answered is: Would your organization rehire Ms. X?

Doing reference checks is time consuming but the information can be valuable. Knowing how people have performed in the past is a guide to how they might perform in the future with their new organization. But it should always be remembered that previous performance of an applicant was conditioned by at least three factors: (1) how they were managed, (2) whether they were in a job that "fit" them, and (3) any personal life factors that may have influenced their performance on the job.

Step 4

Interviewing usually follows after the reference checks have been completed and the applicant seems to be a reasonable possibility for hiring. While reference checks can be difficult to conduct so as to get valuable information, interviewing has even more traps. The ineffectiveness of interviewing usually comes from one of two basic realities: (1) most managers are not good at interviewing, or (2) some interviewees are very good at "reading" what they think will play well with the interviewers. When this happens, the interviewees work at providing what they believe the interviewers want rather than saying what they really think.

A few basic interviewing guidelines can help hiring managers improve the results:

- The interviewer should reduce his or her talking, particularly early in the interview. The interviewer should get the applicants to talk more about themselves and their work history and orientation before interviewees can begin to figure out what the interviewer is seeking. The basic purpose of the interview is to gain insight into what the applicant is like and what he or she has done. In short, the interviewer should "travel around the Model for Growth" as a guide to asking questions about each of the areas of the applicant represented in that model. Later, the interviewer can explain more about the organization, what it is like as a workplace, and what the organization is looking for in the new hire.
- The interviewer should use lots of open-ended questions, especially early in the interview. This also helps the manager avoid legal traps set up by an overuse of specific questions.
  - Example: "What do you believe are characteristics of effective customer service in a retailing environment like ours?"
  - Example: "Tell me about an IT project you managed or helped manage, and explain your specific role."
- Pursue information by following open-ended questions by going at least three layers deep into the topic.

- Example: "You said one characteristic of effective customer service is good credit policy. What do you mean by good credit policy?"
- Example of a second-level question: "Describe how customer recommendations should impact credit policy in our kind of business."
- Example of a third-level question: "In general, how would you provide customers with information on the new credit policy you have just outlined?"

A key to effective interviewing is to use open-ended questions as a source of as much information about the candidate as possible:

- Plan the interview and make sure you have reviewed all reference checks, resumes, and application data before the meeting.
- Asking questions on background — legal ones, of course — provides some information on identity: motivations, basic values, attitudes, and the like.
- Asking questions about recent employment experiences tells more about knowledge and skills, work behavior, and success at achieving work goals. When pursuing these areas, remember to use open-ended questions followed by specific probes to get the details.
- Ask the applicant what information he or she wants about your organization. This is usually best done toward the end of the interview. Remember: a good fit between the applicant and the position is to everyone's benefit. (5)

Because of the ability to see the applicants face to face and pursue information about them with good asking and listening techniques, interviewing is an extremely valuable source of data in hiring. However, even when hiring managers follow guidelines for effective interviewing, like those above, the applicants may "role play" away from who they are toward what they think you are looking for. Back to a slightly modified version of the quote from a CEO business associate used earlier in this chapter: "We conducted an effective interview, I hired him, and his ugly brother showed up."

Step 5

Testing and profiling usually occur when the hiring is getting close to completion. Personality profiling, attitude surveys, and various types of knowledge and skills tests are additional important sources of information for selection. The use of an instrument to measure one or the other of these areas is widespread today after decades of decline in their use, largely because of the fear of lawsuits. (6) The types of instruments available are best understood in terms of what they seek to measure.

1. *Knowledge and skills tests,* sometimes called ability tests, are in abundance for more technical areas. One supplier catalog lists tests in the following areas: office-related tests, computer skill and aptitude tests, and industrial and mechanical tests. For example, an actual test within these categories is the SRA Test of Mechanical Concepts, which purports to measure basic mechanical ability and knowledge of common mechanical tools and devices. (7)
2. *Attitude surveys and interest tests,* which are closely related in what they measure, are also used a good deal by organizations when hiring. The objective here is to measure a significant component of identity and applicant attitudes and beliefs, which are related to behavior and performance. We discussed previously the general relationship between attitudes and behavior. The primary issue is to figure out which attitudes are connected with performance in the organization or position.

   One very useful attitude survey is the Pass III, the Personnel Assessment Selection System. This is a tool that was originally developed in response to the outlawing of polygraphs. This tool, which is completed in about 15 minutes by the job applicant, measures attitudes in three areas thought to be directly related to performance: (a) alienation (attitudes toward work, supervisors, employers, and co-workers), (b) trust (whether the applicant believes in the values of trustworthiness and honesty), and (c) drug- and drinking-related attitudes. An actual example of the use of this tool will help illustrate its benefits.

Organization: Site Oil Company

- Products and services:
  - Convenience food and gas marts
- Markets:
  - The Midwest
  - Florida
  - Automobile drivers
- Structure:
  - Wide geographic spread with its outlet stores
  - Managed by a small corporate staff
  - Active Human Resources function

- Primary issues and problems:
  - Concern with retention of store personnel *and* reduction in theft of their products in the stores

The Human Resources director decided to try the Pass III, along with another potential tool. Being inquisitive and thorough, he sought to correlate both tools to retention of employees. He found the Pass III useful in hiring people who, on average, tended to stay with the company longer than those profiled with the other tool. The director also discovered that the employees with acceptable Pass III scores achieved better performance ratings, on average, from their store managers compared to the other new employees. This appears to be due, at least in part, to the fact that they had fewer "alienated attitudes" about being managed and working for others.

3. *Personality profiling* can be the most beneficial tool for effective hiring because it measures characteristics in the basic identity of the person. At the same time, it is one of the most controversial "testing and profiling" approaches. There is a lot of misunderstanding regarding what personality profiles can and cannot do.

   A concern about personality profiling that still exists, although it is less strong than a decade ago, is the issue of "legality." In general, a valid profile used consistently, with no attempt to discriminate, will not lead to legal damages for organizational users. Courts have recently decided that if an organization shows that its hiring process does not have an "adverse impact on protected groups," then that hiring process will be assumed to be nondiscriminatory. "Adverse impact" is the ratio between applicants and hires for protected and nonprotected groups. In general, the ratio for protected groups should be at least four fifths of the ratio for nonprotected groups to be legal. Of course, anybody can be sued for anything in today's litigious world. The key requirement is to be ethical and nondiscriminatory in hiring so that legal actions are less likely to occur and less likely to lead to adverse decisions against the organization if they do occur. (8)

   Using a valid personality profile consistently for any hiring occurrence can provide useful information on the personality, motivation, and other aspects of the applicant's identity. The concept of validity is somewhat confusing, but essentially there are two important types of validity. First, there is *construct validity,* which essentially is determining that the profile measures what it purports to measure, and that there is a reasonable correlation between what it measures and the behavior of the person completing the profile. There are various ways to validate a profile, including observation of behavior or the correlation of results from the profile being validated with another valid instrument that measures the same behavior. The goal here is to determine if a person is essentially what the profile identifies and see if that person acts as expected based on his or her profile results.

The second type of validity is *concurrent validity.* This has to do with the correlation between an applicant's results on the profile and his or her performance in a *particular job.* There are different levels of concurrent validity. The most general level is when a profile measures how successful people holding a particular position in *different organizations* have scored on a profile. These are national or multi-organizational norms for managers, supervisors, customer service providers, commissioned sales people, computer system installers, etc. The information on how successful persons in one of these job roles have scored on a profile then becomes the benchmark for evaluating profile scores of future applicants for those jobs.

Concurrent validity becomes more concrete when it involves developing preferred scoring ranges for a particular position, or a number of positions, *within one organization.* This can be done in a number of ways, but the most effective way for generating suggested ranges for a job in an organization is conducting a normative study. Normative studies are done by identifying high performers in a position, profiling them, and using those results as benchmarks for profiling scores of future applicants for that position. The results are used as "hire to" profile scores.

There are a large number of personality profiles or "tests" being used in organizations: the Ghiselli Self Description Inventory, Calipers, and the Minnesota Multiphasic Personality Inventory (MMPI), for example. The MMPI is more of a psychological measure, tapping traits such as paranoia and hypochondria. The distinction between psychological tests and personality profiles is important because in most cases organizations are primarily interested in measuring practical aspects of how identity is related to performance, and less interested in knowing about an applicant's psychological abnormalities.

There are a number of factors to consider when deciding which of the many personality profiles to use. These include costs, whether the profile measures practical characteristics related to job performance, construct and concurrent validity, and efficiency of administration. In addition, it would be best if the profile selected is valid for both outside applicants and persons already working in the organization who are being considered for new job assignments. Of all the instruments available, we have found that The Achiever best meets these requirements.

The Achiever measures six mental aptitudes, ten practical personality characteristics, and has two honesty measures. The mental aptitudes include the following: "mental acuity" or quickness to learn, reasoning ability, command of vocabulary, and mechanical interest, among other factors. The personality section of The Achiever has measures of dominance, motivation, extroversion versus introversion, work habits, and similar dimensions.

The Achiever also has two honesty or "faking scales" to determine if the person completing it is being honest in answering questions about his or her personality. Specifically, the instrument measures distortion and equivocation. About ten percent of those completing The Achiever, in our experience, distort or equivocate to the extent that the accuracy of the personality results is in question. By inference, a high distortion or equivocation score probably indicates that the applicant has difficulty with feedback. Receiving feedback is part of most jobs.

As stated previously, finding an effective match between the applicant and the position is beneficial for both the organization and the applicant. Unfortunately, many organizations conduct hiring in such a way that they lose a large number of their new hires within a short period of time. (9) In many cases, poor hiring results from one or the other of two causes: (a) the hiring process has been poorly defined or is not followed, or (b) decisions on hiring are made without sufficient information.

The following case study illustrates more specifics regarding effective hiring and includes one organization's hiring process.

Organization: Ralston Purina Training Function (now Nestle Purina)

- Markets or internal customers:
  - The company provides pet food
  - The training function developed and provided training in many areas, including management, selling skills, and teamwork
- Structure:
  - There were a number of Human Resources Development units at Purina, basically divided by the groups of internal clients they served

- Primary issues and problems:
  - Lacked an effective hiring system: high turnover and not meeting sales goals
  - Needed to hire field sales forces to maximize selling ability and retention of sales people

The training function and the internal clients recognized the need to maximize effectiveness in hiring their field sales force serving the veterinarian and pet specialty stores. Identity factors including personality and attitude directly affect sales performance. In addition, there are specific skills critical to selling success. The Manager of Training and Development for the Pro*Visions division of the company developed the selection process depicted in Figure 8.2.

Purina Selection Process (now Nestle Division), Pro*Visions Department

The training function realized that the performance of the sales force is essential to success of the Pro*Visions unit. The performance of the Pro*Visions unit largely depended on the field sales force selling in sufficient quantities. An internal pre- and post-study of retention demonstrated that both retention and selling volume were significantly better after the installation of the selection process defined above compared to selling performance before the process improvement contained in the new selection process. Retention problems were essentially eliminated. Selling levels increased, and they reached the firm's goals.

OTHER TECHNIQUES FOR EFFECTIVE SELECTION

Because of the importance and complexity of selection, organizations often make use of additional techniques beyond those identified thus far in this chapter. Some of the more permanent techniques include assessment centers, realistic job previews, and various other techniques.

Assessment centers are very common in larger organizations. Depending on the organization, assessment centers perform some combination of the following functions:

1. They are involved in the selection and development of employees with high potential, especially for managerial and executive positions. (10) This focus has resulted in some of these units being called "managerial assessment centers."
2. Assessment centers sometimes function as career development centers, although this is probably less common than the selection

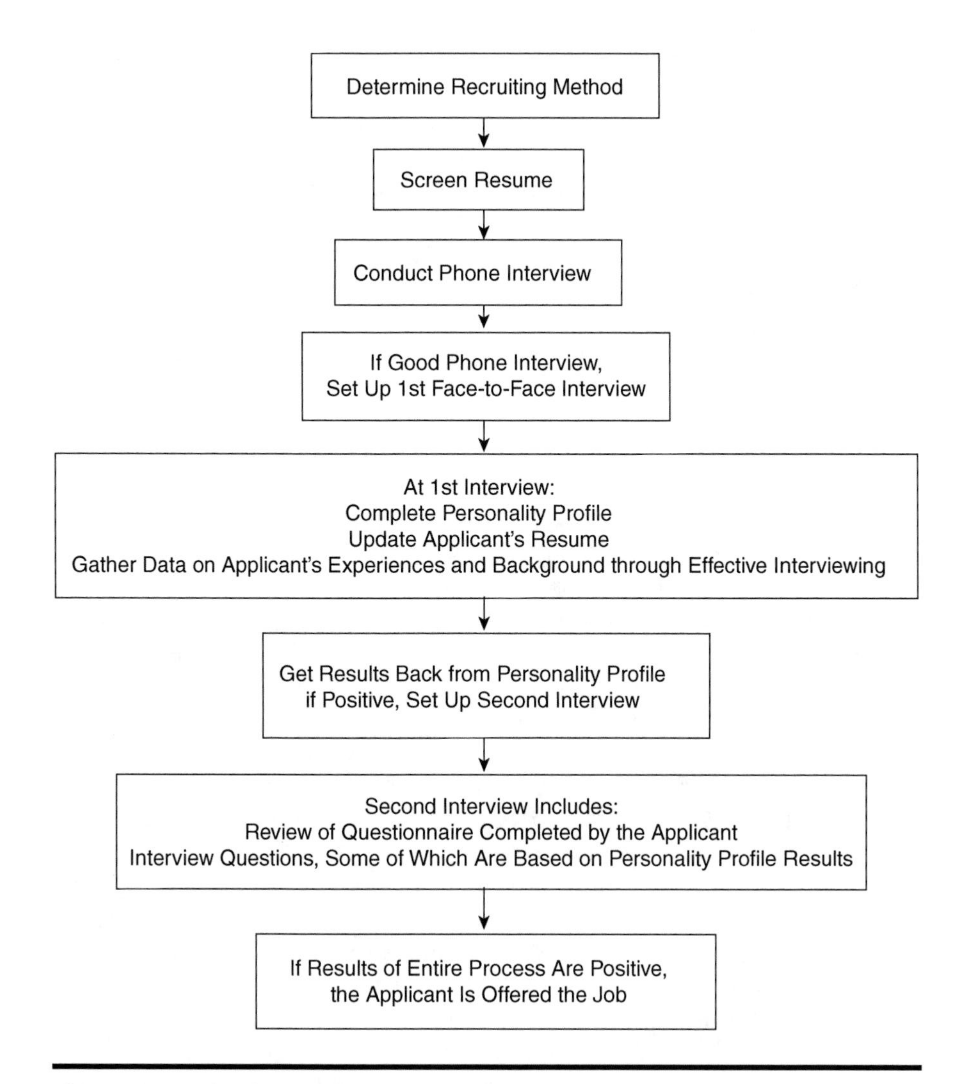

Figure 8.2 Selection process.

focus and executive development function cited above. The essence of career development is to clarify the goals and major identity characteristics of an employee. The objective then is to link the aspirations and characteristics of the employee to the objectives of the organization/team in which the employee works.

3. The third and least common use of assessment centers is in choosing and developing people for new work designs, such as teamwork, project matrix structures, and the current emphasis on self-management. This can include an emphasis on the employee developing and having a strategy for management of his or her own career.

These three primary functions of assessment centers have a common purpose — identifying and improving the fit between candidates and positions. The techniques frequently used by assessment centers include (11):

- "Objective tests" of personality, mental aptitudes, attitudes, and interests
- In-depth interviews
- Individual presentations
- Management games, simulations of work experiences, outbound exercises such as wilderness camping, in-basket activities
- Feedback on leadership style, teamwork skills, and discussion of desirable models for these roles

Assessment centers are, in large part, an attempt to systematize the selection process and make sure it is coordinated and thorough. Many organizations have the rudiments of assessment centers, even if they do not use the organizational label. Selection, in general, and specific approaches such as assessment centers, should be done with an eye toward serving the organization's strategy. More about that in later chapters.

Realistic job previews is a selection technique where the benefits have been well documented. Not surprisingly, many hiring officials having conversations with valued candidates tend to "oversell" their organizations. They do not describe the negative aspects of the job, the team or department, or the organization. They emphasize the positive. This unrealistic perspective has led to occurrences of "job shock" when a new employee starts his or her position and discovers the downside. (12) To overcome this problem, and the resulting poor morale, organizations sometimes structure an activity where employees are introduced to the negatives of the job, such as stress, rapid pace, boredom, long hours, and the like. Among other benefits, this technique appears to reduce turnover.

Various other techniques to improve selection exist. These include an applicant experiencing problem-solving situations with current managers, or spending time with an employee doing the work the applicant will be doing if selected. These techniques all have the same purpose: providing information that will help all involved in deciding if the fit between candidate and position is mutually beneficial.

THREE ADDITIONAL SELECTION ISSUES

1. As stated previously in this chapter, selection officials in organizations frequently do a poor job at making transfer or promotion decisions for current employees. The internal politics of the organization, such

as whether a person is "in" with managers having influence, or whether the current employee has been "loyal" to the leadership, often has undue influence in making promotion decisions. *A candidate from inside the organization should essentially be evaluated the same way as a candidate from the outside.* Proper fit of the inside or outside candidate to the position is the most important consideration. While loyalty and good work of current employees should be rewarded and recognized, it is not in the best interest of good employees to be moved to a position for which they are not suited by personality, attitudes, interests, or motivation. Success in one job within an organization does not guarantee success in a different job in the same organization. Most importantly, poor performance in a job is in no one's interest — not a new employee or one who has been in the organization for many years.

2. As a practical matter, the manager or supervisor responsible for directing the candidate should have at least equal influence in the selection decision. Following the manager of a new hire in the hiring decision is difficult for many executives or business owners. The problem is that either the ego of the executive or the belief that hiring should be the responsibility of leadership. The basic point is that the person who will direct the new hire will feel more responsible for helping that person perform well if he or she has at least equal influence in the selection decision. In addition, they usually know the job being filled better than an executive often two levels above the position being filled. Managers and supervisors of the position being filled often resent having employees imposed on them by their superiors. Even supervisors who are well intended will struggle with this resentment.
3. The selection process is a major opportunity for gathering information that can provide the basis for a number of actions with the candidate once he or she has been selected. Some examples of this valuable information include how to manage the new person, identification of needed training and development, and what information should be emphasized in the orientation. Specifically, the selection process should provide the following information about the selected candidate:
   a. Information from profiling, attitude tests, and interviewing pertaining to the person's identity: personality strengths and challenges, motivation, values, and basic attitudes.
   b. Information from interviews, skills tests, and reference checks to help the organization or team know what knowledge and skills should be provided to the new employee. This should lead to plans for technical training as well as knowledge and skills for

the person's role as manager, sales associate, customer service provider, or team member.

c. Checking references and interviewing, along with feedback from assessment centers, provides information on areas of management style of the new employee if he or she is to manage or supervise. This should result in development plans for the new employee.

One technique that is beneficial to the organization and the person selected is the development of a hiring agreement as a conclusion of the selection process. This contract should include:

- Agreement on basic areas of responsibility and the output requirements for the newly selected person
- Agreement on training and development programs and activities the person is to experience; this should include a listing of the specific concepts and skills that must be mastered by the new hire
- Agreement on the support needed by the selected person to maximize his or her performance, and what his or her manager will do to provide that support

CONCLUSION AND SUMMARY

Selecting the right people for positions in the organization or on a team is key to performance excellence. All the performance improvement interventions discussed in this book are less effective if the persons involved in those efforts are poorly fit for the jobs they hold. On the positive side, the performance interventions described in this book will have a major impact when selection has been well done, and the right people are in the right jobs.

SUGGESTED ACTION STEPS FOR ORGANIZATIONAL OR TEAM LEADERS

1. Use the information in End Note 3 of this chapter to decide on job requirements for areas of hiring or selection needs in your organization.
2. Review the organization's selection process, both for outside applicants and internal applicants.
3. Find or develop methods for effective interviewing and make them available to hiring managers.
4. Review available personality profiles, attitude or interest surveys, and skills tests. Install the appropriate instruments at the proper stages in your selection process.

5. Consider conducting a normative study for the most critical positions in your organization/team. Install the "suggested scoring ranges" in your profiling process that result from that study.
6. Decide if any other selection techniques are needed for improving the probability of getting the right people in the right positions.
7. Make sure that your organization's or team's promotion and transfer process makes systematic use of available data sources.

END NOTES

1. In the strict sense, I do not agree with Collins' argument for "First who…then what?" At least in the broadest terms, the "what" must be outlined before we know what people we need to hire. See Collins, Jim, *Good to Great,* Harper Business, 2001, Chap. 3.
2. Dessler, Gary, *Personnel/Human Resource Management,* Prentice Hall, 1991, p. 172.
3. Aldag, Ramon J. and Stearns, Timothy M., *Management,* College Division, South-Western Publishing, 1991, p. 392.
4. We discuss individual job competencies in Chapter 10. For a thorough discussion of the subject, see Dubois, David D., *Competency-Based Performance Improvement,* HRD Press, 1993.
5. These guidelines are a summary from an AAI training program on hiring.
6. Dessler, Gary, *Personnel/Human Resource Management,* Prentice Hall, 1991, p. 173.
7. SRA Human Resources Assessment Catalogue, p. 4, and the SRA Test of Mechanical Concepts Examiners' Manual.
8. Dessler, Gary, *Personnel/Human Resource Management*, Prentice Hall, 1991, pp. 180–181.
9. Stoner, James A.F. and Freeman, R. Edward, *Management*, Prentice Hall, 1992, p. 562.
10. Cummings, Thomas G. and Worley, Christopher G, *Organizational Development & Change*, South-Western College Publishing, 1997, p. 411.
11. Dessler, Gary, *Personnel/Human Resource Management,* Prentice Hall, 1991, p. 189ff.
12. Aldag, Ramon J. and Stearns, Timothy M., Management, College Division, South-Western Publishing Co., 1991, p. 306; Cummings, Thomas G. and Worley, Christopher G, *Organizational Development & Change*, South-Western College Publishing, 1997, p. 409; and Dessler, Gary, Personnel/Human Resource Management, Prentice Hall, 1991, pp. 252–253.

Chapter 9

LEADERSHIP, MANAGEMENT, AND TEAMWORK

INTRODUCTION AND LINKAGE

Previous Chapters as a Linkage to Leadership, Management, and Teamwork

This book is based on the belief that we know the characteristics of an effectively performing organization of whatever type: for-profit or not-for profit; service or manufacturing; or any type of business or industry. This belief is supported, as we see it, by substantial research as well as more than 34 years of experience.

An effectively performing organization is one that defines and achieves worthwhile objectives that help the organization define and achieve success. Thus far, we have said that this successful organization has the following characteristics:

1. It has defined a specific strategy including the following: a vision of its desired and intended future in detail, a detailed listing of its current strengths and weaknesses, and a list of annual strategic objectives that provides the organization with goals for performance success and improvement.
2. The organization has a clear definition of its critical performance areas and assesses performance successes and gaps in those areas (key performance indicators, KPIs).
3. The organization has established goals and related KPIs throughout the organization. It has goal setting at every level and in every department and unit comprising it.

4. The organization has maximized its attention to the importance of learning and has set up, within its capacity, learning opportunities for all employees.
5. It has a hiring process that is directly connected to the organization's strategic plan and makes the best use of effective hiring techniques.

All of these characteristics of an effectively performing organization require the intense and persistent dedication of top managers who set the directions listed above: commitment to the development of a clearly defined strategy that is pursued by everyone, making goals a way of organizational life, maximizing effective hiring, etc. Then it is the role of middle managers, supervisors, and team leads to carry the directions set by leadership to the employees and get the work done.

The most powerful CEO or president leading an organization cannot be successful by himself. It takes other managers, supervisors, and all the employees working together toward common objectives and goals for successful performance. Getting motivation and dedication for successful performance throughout the organization requires effective leadership, which is primarily the leaders' ability to communicate clear direction and obtain the motivation and dedication of the people in the organization or team (department).

1. The leader of the organization or team sets the tone for when and how teamwork is used and whether it is used sufficiently. While it is true that leaders often make the mistake of discouraging teamwork when it could increase quality and acceptance of decisions, the opposite is also true. Some leadership over-emphasizes and over-uses teams. In this case, the frequency of "meetings" and team involvement in areas where they do not belong end up interfering with performance. Leaders need to know when to use teamwork and when to not use it. Guidelines for using teams are discussed more fully in Chapter 10.
2. The attitude team leaders have about the ability of their team to perform is probably the most critical factor in actual team performance. To slightly amend an ancient phrase, team members might say the following to their leader: "If you say we can or say we cannot (perform), either way you are right."
3. Leaders need to make sure that teams have a specific strategic vision, clear goals, and a defined purpose. Sometimes a team depends on the specific strategic vision of the larger organization in which they work. Production teams are an example of groups that have their strategic direction handed to them. Another example is sales teams, which often carry out the marketing strategy

of their organization. But even where the basic products and services the team provides are determined by others, the team needs goals, leadership direction, and standards and procedures guiding their performance. Teams also need techniques for diagnosing their current performance and targeting performance improvement interventions.

4. Leaders of organizations or teams directly affect performance of their teams in a number of ways:
   a. How they recognize and provide feedback on performance
   b. How they define their role and that of managers under them in terms of developing subordinates
   c. How "mistakes" are handled

Leaders often demonstrate two very different negative attitudes that negatively impact team performance. In some cases, poor performance is expected and tolerated, and there is little belief in performance improvement or attempts to develop people. In another leadership direction, any performance that is not exactly at the level expected by leadership is punished. In the first case, people are not motivated to get better at what they do. In the second case, people are afraid to be anything less than perfect as defined by the leadership. Learning from mistakes is not possible because mistakes are not tolerated. Therefore, innovation is stifled.

Neither of these common leadership styles is productive for effective team effort.

THE IMPORTANCE OF LEADERSHIP AND TEAM MANAGEMENT

In recent years, leadership development has become one of the central themes in management literature and in performance improvement initiatives in organizations. In its 1998 report on training in organizations, the American Society for Training and Development (ASTD) said that 93 percent of all organizations offered courses in Management–Supervisory skills. (1) The ASTD's 2005 report noted that close to 13 percent of the training "learning content" of its sampled group was "managerial and supervisory."

These training programs provided in organizations of all types frequently offer leadership or managerial suggestions and techniques for people at various levels of the organizational structure. In the 1998 ASTD report, 63 percent of the organizations had Executive Development Programs, undoubtedly having major leadership components.

Further evidence of the popularity of the leadership topic is the enormous number of books available today on leadership vision and the changing role of managers. Some of these publications have been referenced in previous chapters. However, a good deal of confusion exists about the topic of leadership, despite its popularity. First, the terms "leadership" and "management" tend to be used interchangeably, although it is useful to distinguish these two areas of knowledge and skills. In general, *management is the performance of the long-recognized management functions: planning, organizing, directing and evaluating*. Staffing (selecting people to fill positions) and representing the team should also be added to the list of management functions.

Leadership is the use of influence in encouraging, directing, and coordinating persons toward the accomplishment of the organizational or team strategy and goals. People who have the qualities or characteristics to be able to exert this influence are called "leaders," or effective leaders. (2) In a sense, "ineffective leadership" is an oxymoron, although that phrase is used when people are talking about those having formal positions of leadership, such as a CEO, but not being good at the use of leadership influence.

As discussed in Part I of this book, the first requirement of leaders is making sure that the thorough vision of the desired performance of the organization or team is defined and communicated. Then leadership needs to provide the supporting systems and resources, including knowledgeable, skilled, and motivated personnel, to get the work done for achieving the strategic vision and the supporting goals and standards. At this point, being good at the management functions is important in creating the performance necessary for achieving leadership's strategic vision.

Ideally, therefore, those heading organizations and teams should be effective in both management techniques and functions and in leadership. People good at managing but not leading have trouble getting people to follow their plans, organizing efforts, directions, and evaluations. Those having great ability in developing and communicating a strategic vision, but little ability in, or patience for, managing get a lot of people inspired but comparatively little actual performance occurs. In fact, many heads of organizations or teams are better at managing than leading, or vice versa. The good news is that the ability to lead and to manage can be learned and developed.

SOURCES OF AND CHANGES IN AN INDIVIDUAL'S LEADERSHIP AND MANAGEMENT BEHAVIOR

The Model for Growth (Figure 9.1) discussed previously helps clarify the sources of a person's leadership or management style. For that matter, this extremely useful model helps us understand our performance, or the performance of others, in many different life roles.

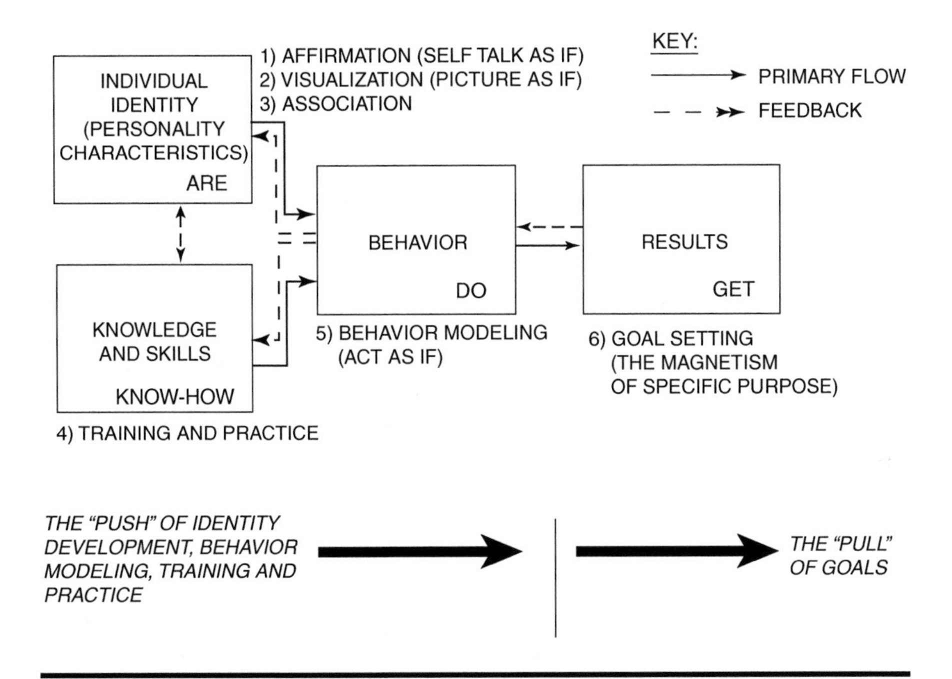

Figure 9.1 Model for Growth.

Readers may remember from previous discussions that the first box, "Identity," represents who an individual is as a person. This includes personality, motivations, attitudes and values, and other parts of the human being's identity. Some portion of our identity comes from heredity; how much is continuously debated by psychologists and other students of the human condition. Some parts of our identity come from conditioning or experiences we have had in life. Conditioning means that our identity is partly what we have done and where we have been. And, as has been said by many personal development experts, we are also right now "becoming" what we will be in the future. If our identity to some extent changes with our experiences, then it is logical to believe that when we change our experiences, we will, over time, also change our identity. This is true of any type of experiences we have, including those of leading and managing. In summary of this major point, we can re-condition ourselves.

We have had experiences with a number of organizational leaders who are pessimistic about people changing what they are like and how they behave. One version of this pessimism is the phrase "People don't change." Of course we all change. The question is: can we influence how we change, and in what direction our changes occur? Or, fatalistically, are the changes that we experience beyond our influence?

One source book in particular provides great insight into the issue of change and how it occurs. "In summary, development (of humans) is a multifaceted and complex process, involving gains and losses, growth and aging, and more, brought about by both maturation and learning" (Insert is ours). (3)

We also know from the Model for Growth that our behavior not only is influenced by our identity, but is also influenced by our knowledge and skills, which is the "learning" described in the above quote. This learning includes our knowledge and skills in leadership and management as well as in the rest of life. That is rationale for the billions of dollars spent on leadership and management courses offered by organizations. The objective is to change the performance (behavior) of managers and leaders to become more effective in their professional roles.

It seems clear that we all change over time. The primary question is the extent to which people can manage their own change, and how that can best be done. "Growth" is desirable change over which the individual has influence. The Model for Growth says that if a change in behavior results in reaching goals (improved performance), then the change will become permanent because individuals like the improved results. Their new behavior and the results they get will be rewarding and satisfying. Their tendency will be to continue the new behavior and make use of the new knowledge and skills.

Two very practical examples from today's work world will be helpful for clarification here. Today, most of us have the need to be at least minimally sufficient with computer operations. Writing this book within the deadline would be difficult if word processing and various graphic technologies were not available. Just a few of the technical techniques that are required in preparing and sending this book to the printer include spell check, spreadsheet use, e-mail, and graphic design and editing techniques. These technology procedures did not exist when I went to school. They had to be learned.

So, this would infer that people can and do develop knowledge and skills about technology. What about other areas, such as the "soft skill areas" of communication or managing? Many times, our clients ask how to better motivate their employees. This is a complex topic, discussed later in this chapter. As an example here, we might make a key point that people are most motivated to work on a project on which they have some input on the basic decisions about that project. Therefore, the technique we teach team leaders or managers is to get others on the team to discuss their ideas before the manager starts talking about what he or she thinks about the project. When managers do this, they avoid having most of the discussion focused on their thoughts, which tends to result in less motivation by people who had less input. While it takes time and effort, people can learn to listen and ask questions first and speak second when they are the people with the "power."

So our *identity,* especially attitudes and personality, and our knowledge and skills drive our behavior, and we can change our behavior best if we see better results from the behavior changes. These kinds of changes occur for any life area, including our behavior as managers and leaders. But so does the behavior of others who have power or influence over us. This is "Behavior Modeling" as described in the Model for Growth. Most managers and leaders who are not at the top of the organization report feeling that "what is accepted management style" in the culture, or by their own boss, affects their behavior, making them behave differently than they would like. This is "role playing," which is managing our behavior away from a natural expression of our *identity* and *knowledge and skills* toward what is expected. Role playing can have desirable or undesirable results. Whatever the results, role playing is stressful because it takes us away from our natural identity and requires constant self-management.

If role playing leads to more effective performance as a manager or leader, then it is desirable, although it is usually stressful and uncomfortable. The key question, long a thorny issue in management and leadership discussions, is whether there is a model of management and leadership worth following as individuals work at growing their own performance as a manager, supervisor, or team leader.

IMPORTANCE OF LEADERSHIP AND MANAGEMENT IN ORGANIZATIONAL OR TEAM PERFORMANCE REVISITED

One way to understand the importance of management and leadership in the performance of organizations and teams is to refer to the Model for Organizational Success (Figure 9.2).

As a brief review, remember that the *strategy* component of an organization's success is clear statements about the following: What products and services do we provide? What markets do we serve? What methods of marketing and distribution do we use? This is where leadership must first perform effectively if the organization is to be successful. The need is to develop a detailed vision of this part of organizational success and then communicate it to the organization or team.

The *operations* of an organization include clear decisions and policies on topics such as the following: How do we hire, and who are we looking for? How do we manage our people? What technology do we have and need? How do we recognize and reward? How do we evaluate and make needed changes in our products and services? The definition of how these things should be done is the responsibility of leadership. The specific initiatives, programs, and daily leadership to make sure

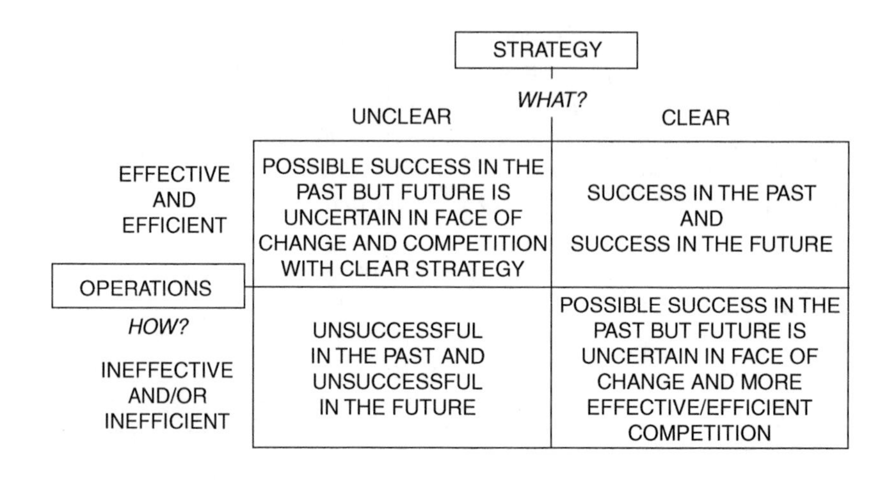

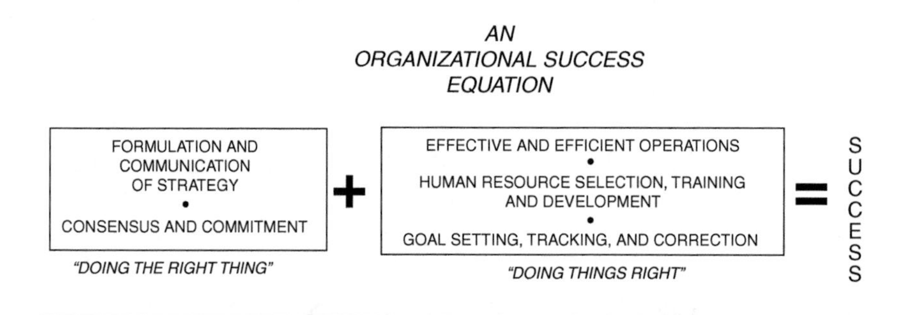

Figure 9.2 Model for Organizational Success. (Acknowledgment to Tregoe, B. and Zimmerman, J., Top Management Strategy, Simon and Schuster, 1980.)

the operations are effective and efficient are largely the responsibility of management.

Somehow, therefore, the establishment of a clear strategic vision and the management of daily operations both must be done for an organization to be successful. It is our impression from more than three decades of organizational work that most organizational or team managers are generally better at managing operations than they are at leading their team to strategic achievement. In the case of a leader who has trouble with the strategic vision requirements of leadership, using a top team to develop that vision is even more crucial to organizational success. Those who are good at building and communicating a strategic vision, but less capable at daily management, need a team of managers who help them with daily operations. Occasionally, a person becomes effective at both strategic leadership and operational management, and has a major role in both strategy and operations. This person is extremely valuable to his or her organization.

MODEL OF EFFECTIVE MANAGEMENT AND LEADERSHIP PERFORMANCE

What style of leadership or management is most effective at accomplishing the multiple roles described above: defining and communicating the strategic vision, making sure that the systems are there to make strategy happen, and helping ensure that daily operations are effective and efficient? For much of the history of the writings in Western civilization, this has been a central concern and area of disagreement. Plato gave us *The Republic*. Machiavelli wrote *The Prince,* which is where the word "Machiavellian" originated. Many authors in the 19th and early 20th centuries discussed the "Captains of Industry" who often sacrificed everything to growing their business at whatever cost. In more recent times, the literature on leadership and management has included a focus on contingency management and "situation leadership." Situational leadership, still very popular today, is succinctly described in the following quote.

> "According to Situational Leadership, there is no one 'best' way to go about influencing people. Which leadership style a person should use with individuals or groups depends on the readiness level of the people the leader is attempting to influence." (4)

Situational leadership rejected what has been called "normative approaches," the belief that there is a best way to manage and lead. After centuries of disagreement over the best leadership approach, moving to approaches that were more "adaptive" to people and situations was a natural occurrence. In the 1970s, however, a series of studies was begun that led to a data-based model of "the best" management and leadership style. The Achieving Manager Project, based on studies of achievement with thousands of managers over three decades, resulted in a model of effective leadership/management: the High Achieving Manager Model. (5)

There were two primary issues to answer in this gigantic study of managers. First, once and for all, can we find some way to define and measure leadership and management effectiveness so that one group of leaders can be identified as more effective than other groups? That is, can we identify successful leadership and management performance in an applied way so that we can measure it in the real world of organizations?

The second issue had to do with figuring out what makes successful leaders and managers more effective than others. If we can find a group of High Achieving Managers, what do they do to be high achievers compared to other leaders and managers?

Defining high achievers in organizations was a difficult task for the Achieving Manager Project. What was needed was a measure of achievement

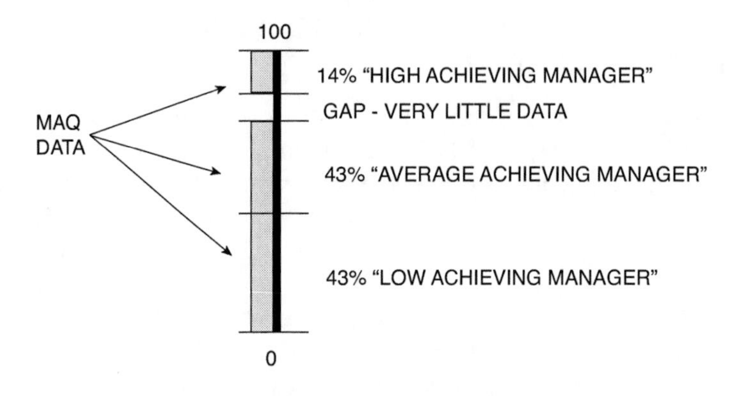

Figure 9.3 Managerial Achievement Quotient MAQ.

that was objective in the sense that all could look at it and at least generally agree on the results of the measurement when it was used to assess actual managers. In addition, the project director, Dr. Jay Hall, believed it was necessary to use measures of achievement that were important to actual managers. They decided to use the Managerial Achievement Quotient (MAQ) for distinguishing the performance of managers. Specifically, they developed individual scores on achievement for each manager in the study by giving him or her points for his or her success in the organization.

Here is how it worked. The original study in the 1970s was performed using about 50 organizations with more than 12,000 managers. The achievement of each manager was rated based on a formula measuring his or her promotion rate compared to his or her age. Because promotion usually involves an increase in salary and the number of the manager's direct reports, promotions also indirectly measure other achievement factors. Remember that the study covered thousands of managers from many different types of organizations. Using success in promotion rate compared to others in the same organization, the results were as shown in Figure 9.3.

All managers were placed on the above curve based on their numerical scores from the MAQ rating their promotion rates in relationship to their ages. The results were dramatic and reflected a standard bell-shaped curve, also known as a standard distribution. In the 1970s version of the study, the High Achieving Managers, those being promoted more rapidly than others, equaled between 14 and 16 percent of all managers. At approximately the 16 percent range for Achieving Managers, the "Average Achieving Manager" numbers began to increase. Thus, the largest percentage of managers in terms of promotion rate was about 70 percent, who were average or below in achievement as measured by promotion rate. These managers are represented as being in the middle of the bell-shaped curve. Following the Average Achieving Managers was a small percentage of

Low Achieving Managers, equal to roughly the same percentage as High Achievers. So, some 2240 to 2560 managers (14 to 16 percent of the total sample of 16,000) were High Achievers as defined by promotion rate and the related growth in power and rewards. The remaining managers were average achieving and below. In the past few years, additional feedback from managers completing follow-up studies has shown a slight tendency to fill in the gap between the High Achieving Managers and all others. That seems to imply that some learning about good management is occurring, although it is minimal.

From a research perspective, this rating scale approach is not perfect. Some managers and leaders, when hearing about this study, ask why Hall and his team did not directly measure the achievement of organizational goals or some type of financial contribution to the organization's success. First, both of these measures (goals achievement and financial contribution) have their own methodological problems. Others reviewing this study are concerned that company politics and nepotism could have influenced the results. The basic point is that with thousands of managers from a large number of very different organizations, we find that a small percentage of managers are recognized and rewarded by their organizations more than other managers. The basic question is why they had this success.

When discussing this part of the Achieving Manager study, organizational or team managers are usually impressed with the magnitude of the sample group and the potential benefit of finding a model of leadership based on what thousands of high achieving managers do. While some reservations about the methodology are reasonable, biased politics in individual organizations would be more significant if the scope of the study was a single company or a few organizations. Also, concerns about bias would be more convincing if this were a very small sample of managers, instead of 16,000. It is reasonable to assume that the size of the sample of managers and the fact that they are from many different organizations means that many elements were involved in their success, including outstanding performance. A significant number of managers were clearly more successful in promotion rate than others. It is reasonable to infer that some part of their success was related to both performance and non-performance factors. And, of course, company politics probably had some influence on the promotion rate of the average and low achieving managers as well.

Even if the MAQ does measure performance achievement by measuring promotion rate, why not directly measure goal achievement or use the performance evaluation system in the organizations? The problem is that many organizations have only fuzzy or ill-defined goals, so how can one know if they have been achieved? In addition, many organizations either do not have performance evaluation systems or they are vague in what they measure, and are themselves very subjective and political.

If one is willing to accept the notion that over a large sample of managers from many different organizations, promotion rate in relationship to age is a measure of performance, then the model becomes a powerful one. Two basic considerations are usually the greatest factors in accepting the Achieving Manager Project as valid. First, the same methodology was used again in the 1980s with thousands of different managers and different organizations than those used in the 1970s version of the study. The results were largely the same — a small group of High Achieving Managers who were being promoted more rapidly than others. Finally, on this point, the study was replicated one more time in the 1990s, again with a large number of managers from different organizations. The results of this third version of the study were essentially the same as in the two previous studies. In these three decades, there was a percentage of thousands of managers who were able to do whatever it took to achieve promotion rates more rapidly than in excess of 80 percent of all the other managers in the sample.

One major reason why most organizational/team leaders introduced to the High Achieving Manager project accept the validity of the model is this: When the study measured how the total sample of managers worked with their direct reports, the results were dramatically different for High Achieving Managers compared to low and average achievers. This difference in how high achievers managed and led their direct reports was demonstrated in the 1970s study and repeated in this project in the 1980s and the 1990s. High Achieving Managers behave differently than other managers when working with their direct reports. While there is no direct evidence that their style of leading or managing itself led to the comparatively high promotion rate of this small percentage of managers, it clearly did not stop their success in climbing the organizational ladder.

Let us return to the Model for Growth (Figure 9.4) to review why the Achieving Manager Project is so important.

One method by which people can *grow* their performance in any area of their lives is to find a model in that area that they believe is valid and effective. We are all influenced by models throughout our lives. This is especially true when we are inexperienced in some area and have not developed a style. So, new employees are especially susceptible to "learning" management style from those they see in their new organization. They will then accept one model, or elements from a number of leadership and management models, which they will come to use when they themselves become managers. Why people accept one model over others is complicated but undoubtedly has to do with values, attitudes, personality factors, and how much of one managerial or leadership approach they see compared to others (conditioning).

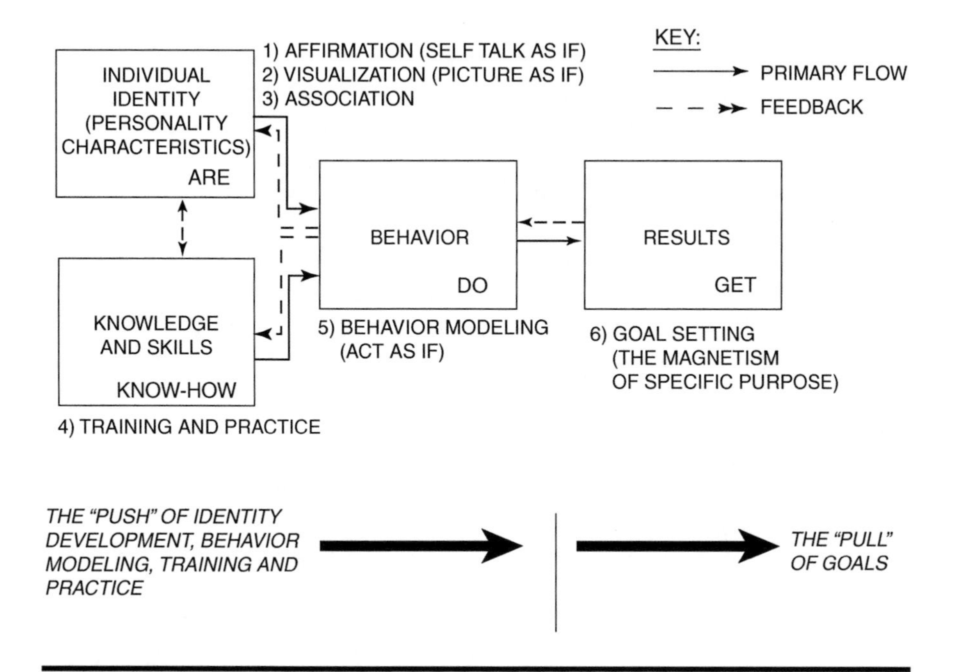

Figure 9.4 The Model for Growth.

Therefore, accepting a model of management and leadership as valid means that *leaders and managers can choose to try and become more like the manager they believe in most.* They can use that model to directly change their behavior. "The key to becoming the Achieving Manager is to learn to act like one."

So what did these High Achieving Managers do differently than the Low and Average Achieving Managers? To answer this major issue, the research team looked at six areas of management and leadership. These areas of management style were as follows:

1. Management values (beliefs about people generally and workers specifically)
2. Involvement, access, and participation (now called participative management)
3. Communication style
4. How managers seek to motivate others and what motivates them
5. Power and empower (how they use power)
6. Overall leadership style

In all three Achieving Manager Projects, the research team adopted models of behavior, beliefs, and attitudes in each of these six areas of management and leadership. They then used validated surveys to get feedback from three direct reports of each of the 16,000 managers. Thus, each manager was described by three direct reports in terms of how he or she behaved in each of the six areas listed above using the models Dr. Hall selected describing management and leadership.

The results demonstrated that the High Achieving Managers as a group tended to behave similarly in each of the six areas of leadership and management listed above. As a group, the High Achieving Managers were described by their direct reports as behaving similarly in the six areas of leadership and management and very differently from the 84 percent of low achievers and average achievers. Keep in mind that these managers and their direct reports came from many different organizations. The input from direct reports was about the individual managers, but the results showed that as a group they were similar in the way they managed and led. They also were receiving rapid advancement in their organizations compared to the other 13,400 managers in the study (84 percent). In what ways do the High Achieving Managers provide leadership and management as shown in the feedback from their direct reports in each of the six areas studied?

MANAGING AND LEADING AREA 1: MANAGEMENT VALUES (BELIEFS ABOUT PEOPLE AND WORKERS)

In measuring manager values about people, the High Achieving Manager Project used a values and attitudes model from a classic work by McGregor, "The Human Side of Enterprise" (Figure 9.5).

Direct reports from the High Achieving Manager Project completed instruments that measured how their managers treated them. They were *not* directly evaluating their managers on X and Y, and the surveys used (Teleometrics Instruments) did not mention the X/Y Model in any section of the survey seen by the direct report as they described their managers. However, the survey instruments led to scores on X and Y for each of the managers and identified how much X versus Y they exhibited.

The graphic in Figure 9.6 shows how the X and Y results were significantly different for the High, Average, and Low Achieving Managers.

The implications for these results come from the "Pygmalion effect," known in more common language as the self-fulfilling prophecy. This is the idea that we get what we look for and what we expect from others, including those who work for us. Keep in mind, for example, that one of the most significant factors in the performance of teams are the expectations about the team held by the leader. That is also true of individual performance. Managers with a lot of Theory X in their beliefs about workers expect people, on the whole, to dislike and want to avoid work and to need to be controlled if any performance is to occur. Theory

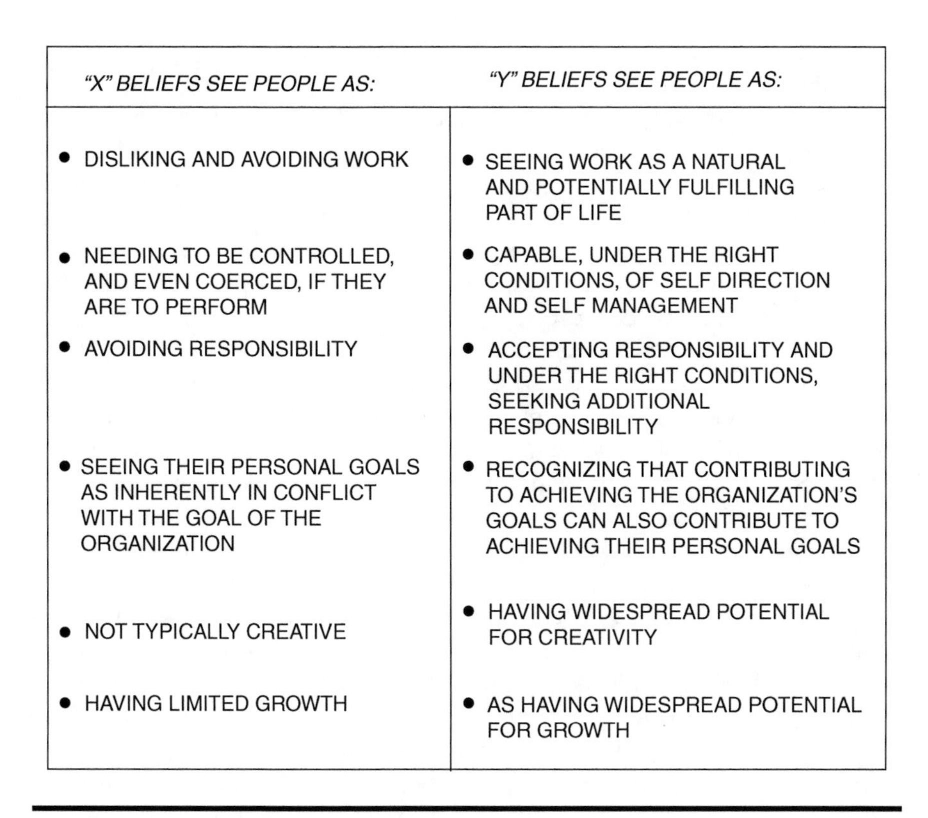

"X" BELIEFS SEE PEOPLE AS:	*"Y" BELIEFS SEE PEOPLE AS:*
• DISLIKING AND AVOIDING WORK	• SEEING WORK AS A NATURAL AND POTENTIALLY FULFILLING PART OF LIFE
• NEEDING TO BE CONTROLLED, AND EVEN COERCED, IF THEY ARE TO PERFORM	• CAPABLE, UNDER THE RIGHT CONDITIONS, OF SELF DIRECTION AND SELF MANAGEMENT
• AVOIDING RESPONSIBILITY	• ACCEPTING RESPONSIBILITY AND UNDER THE RIGHT CONDITIONS, SEEKING ADDITIONAL RESPONSIBILITY
• SEEING THEIR PERSONAL GOALS AS INHERENTLY IN CONFLICT WITH THE GOAL OF THE ORGANIZATION	• RECOGNIZING THAT CONTRIBUTING TO ACHIEVING THE ORGANIZATION'S GOALS CAN ALSO CONTRIBUTE TO ACHIEVING THEIR PERSONAL GOALS
• NOT TYPICALLY CREATIVE	• HAVING WIDESPREAD POTENTIAL FOR CREATIVITY
• HAVING LIMITED GROWTH	• AS HAVING WIDESPREAD POTENTIAL FOR GROWTH

Figure 9.5 McGregor's Theory X and Theory Y.

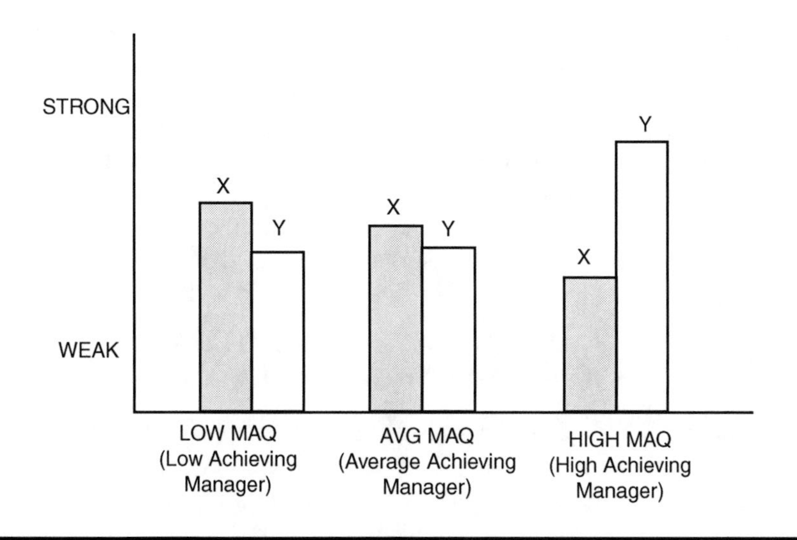

Figure 9.6 Theory X and Theory Y results.

Y managers see people as believing work is natural and having widespread potential for growth. Perhaps the most significant factor in the X/Y Model is that the High Achieving Manager believes that workers often work to help the organization achieve its goals because they see organizational success as helping them achieve their own individual goals. Motivation comes primarily from the individuals' expectation that if they help the organization or team become successful, they will get their needs met. This association or direct tie between organizational goals and individual goals is a primary objective of the basic approach recommended by this book: *organizational development.*

In the world of U.S. organizations, being Theory X does not increase promotion opportunities for managers; rather, it decreases advancement possibilities in most organizations. The Achieving Manager Project and our three decades of experience in organizational development support this principle. Clearly, there are exceptions to this rule, and many managers operating in fear-oriented organizational cultures seem to act as if their nastiness toward their employees guarantees their own success. But fear is a poor motivator and has short-term influence over performance. Attempts at managing others by making them fearful also often results in employees finding ways to get back at the "bully."

It is important to note that the High Achieving Manager is not a "non-manager" or a "wimp" who lets poor performance slide. These managers expect effort and hard work to achieve results. They provide support, clear direction, feedback on performance, and look for opportunities to mentor where they can. *High Achieving Managers expect performance.*

Not only did the three direct reports describe the behavior of their managers using the X/Y Model, but they also described their own preferences for X versus Y treatment from their manager. In addition, the direct reports for each manager described how much X versus Y they believed they received from their manager compared to what they wanted. Keep in mind that this survey completed by the direct reports did not identify the X/Y Model as the basis of what the survey questions sought to measure. This behavioral focus with no mention of the basic models behind the survey questions was true for most of the Teleometrics surveys used in this project.

MANAGING AND LEADING AREA 2: PARTICIPATION, ACCESS, AND INVOLVEMENT (PARTICIPATIVE MANAGEMENT)

The Achieving Manager Project also generated feedback for managers on the uses of participative management techniques. In participative management, "the emphasis is on joint decision making about events that have future implications for the parties involved, and over which they can realistically exert influence." (6) It turned out that High Achieving Managers

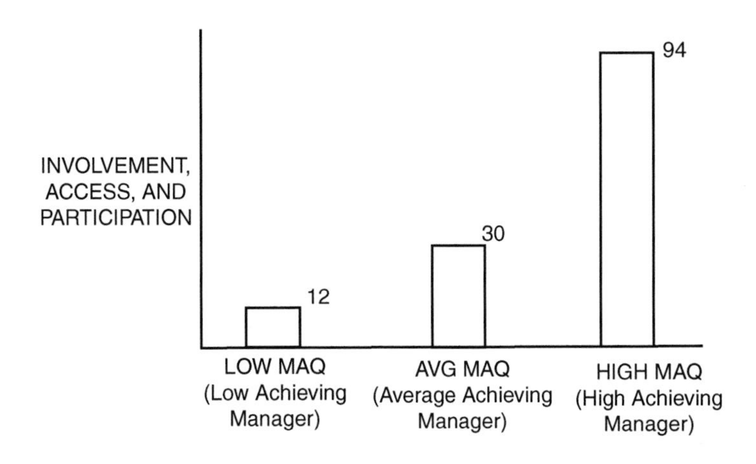

Figure 9.7 Participation and involvement.

made more use of participative decision making than the other 84 percent of the managers in the study (Figure 9.7).

Remember from the discussion earlier that teamwork is not needed in all settings for all decisions. Teamwork is beneficial when you need both *quality* decisions and *acceptance* of those decisions. The logic of using quality and acceptance is strong because it maximizes your effective use of teams and helps you decide when not to use them.

The results discussed under X/Y earlier in this chapter demonstrated that High Achieving Managers expect performance from employees and believe in those employees' motivation to do good work and their ability to learn. Remember that these High Achieving Managers are also rewarded by their organizations through a rapid promotion rate and related increases in influence. Now we see that High Achieving Managers also use participation as an important part of their style, especially when they are looking for quality decisions and acceptance of those decisions. Management values and participative management are connected. If managers have strong expectations for the performance of their workers, they are more apt to use participation to get input from their people because they value it. If managers have negative expectations about their workers, as was shown to be true of Low Achieving Managers, why have the workers involved in deciding anything? The managers do not value their contribution.

MANAGING AND LEADING AREA 3: INTERPERSONAL COMPETENCE AND COMMUNICATION STYLE

Communication is one of the areas viewed by many organizational and team leaders and employees as one of the most important dynamics in

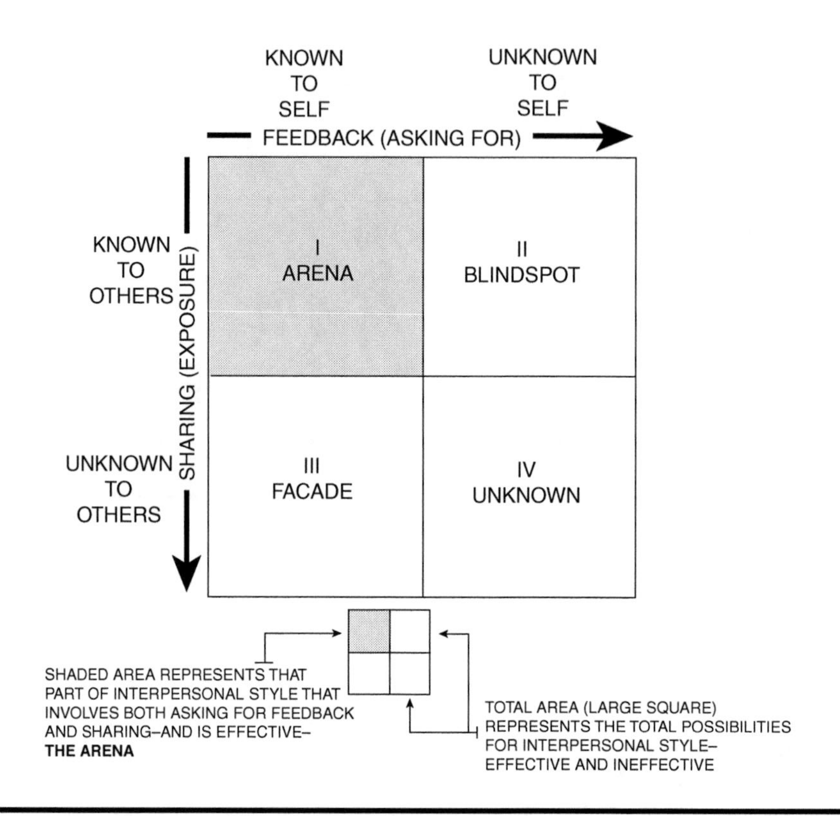

Figure 9.8 Johari Window. (Concept developed by Joseph Luft and Harry Ingham; Luft., J., *Of Human Interaction,* National Press, 1969.)

organizational performance. Communication problems is one of the most frequent areas identified in our surveys of organizational cultures as well as in organizational development and managerial literature. The Achieving Manager Project staff had the insight to include communication as part of the assessment of managers. This is understandable because ineffective communication is often a source of organizational problems. Further, effective communication is necessary for performance improvement to occur. The Johari Window (Figure 9.8) is the model used in the Achieving Manager Project.

The Johari Window focuses on the importance of two dimensions in communication: (1) asking for feedback and (2) sharing information (exposure). When people do both, they have an *arena* "where the action is." When they do not ask for feedback, they are operating without information, and so they have a *blind spot.* When they do not share information and feelings, they are keeping things to themselves, so that is their *facade.* There are four basic styles of communication resulting

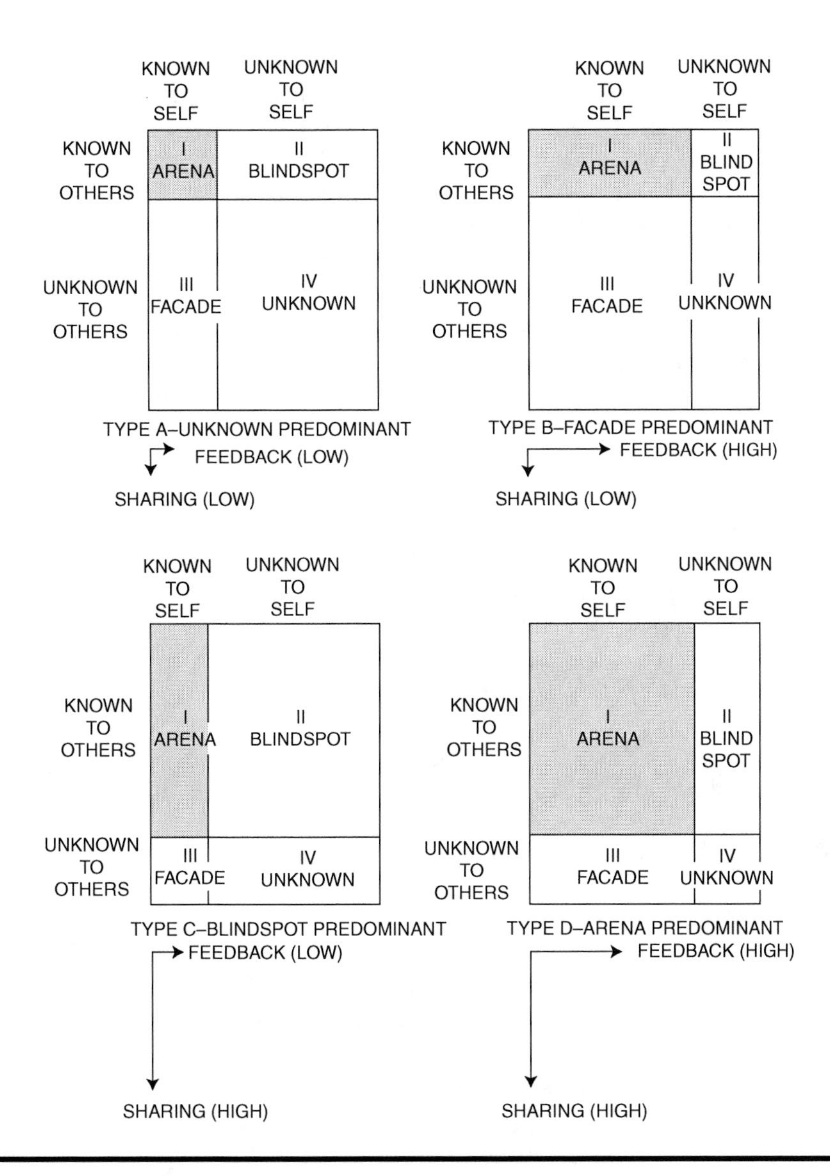

Figure 9.9 Johari Window communication styles.

from how good a person is in asking for feedback and sharing information. Those styles are presented in Figure 9.9.

The Type A style is very closed, with little effort at getting feedback and little sharing of information. This style of communication is troublesome in any situation requiring coordination or teamwork between people. In our experience, the Type A style is often a person who is shy, lacks confidence, or does not like to interact with others. Many factors can cause a person to be "reserved," unwilling to share much information with others and not particularly interested in hearing from others. Leaders and managers must be willing to communicate with others to provide directions, coach and counsel, give feedback on performance, etc.

The Type B style of communication involves a lot of asking for information or feedback, and little sharing. The popular television detective from a number of years ago, Columbo, provided an entertaining version of the Type B communication style. He kept asking questions, without telling what he knew or thought about the suspected killer, until he got his criminal. The problem with the Type B style of communication is that people feel "interrogated" when the person with this style only asks questions and does not share any information. Fear, especially if the Type B person has positional power, is natural and a shutdown in the conversation usually occurs.

The Type C style of communication is frequent among top-level executives, especially in the business world. Type C communicators tell people more than they ever need to know but fail at feedback reception. They do not listen much, but they tell a lot. An excessive and ongoing use of this style is often a reflection of arrogance and overuse of positional power. Sometimes, this style of "tell but don't ask" is reflective of an individual's highly assertive identity. This style of communication has two major drawbacks. First, because the Type C communicator does not listen much to others, they do not learn what other people think or need to know. Second, this type of communicator does not gain much buy-in to their ideas because the person or persons they are communicating with are given very little chance to contribute to the topic under discussion.

The Type D communicator is someone who both listens to others and shares information as needed. This style is essential to teamwork and collaboration. This is the most effective style for managers, supervisors, and team leaders. However, most people more often use Types A, B, or C. There are many reasons for our communication style, including habits learned from childhood and what our parents encouraged, our personality and motivation, and the culture of the organization in which we are working. (7)

The reader has probably figured out by now that the High Achieving Managers were shown to be primarily Type D communicators in the Johari Window assessment provided by their direct reports. Again, the direct reports simply described behavior, and the results of those descriptions led to a Johari Window score. The direct reports did not "evaluate" their managers by comparing them to the Johari Window; rather, their descriptions of the manager led to the score. The High Achieving Manager was 25 percent better than the Average Achieving Manager in both feedback and sharing information. The Average Achieving Managers tended to score like the average person in our society in communication effectiveness.

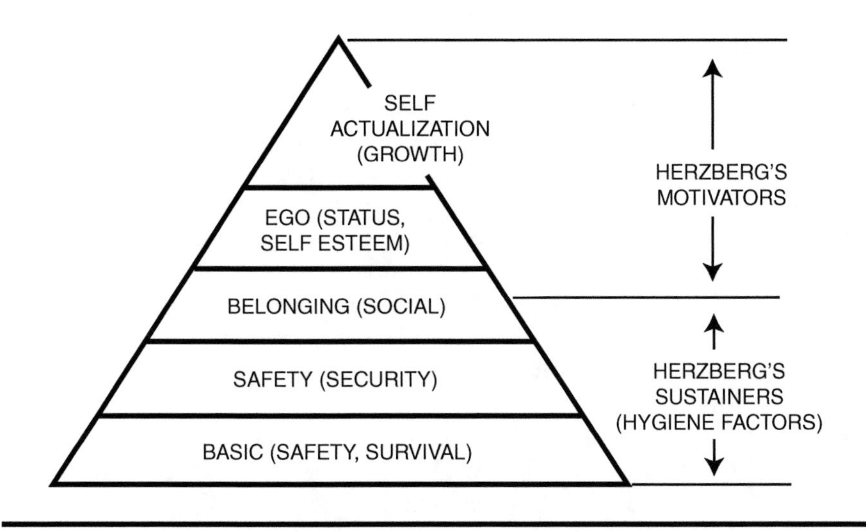

Figure 9.10 Maslow's hierarchy of needs. Motivation illustration.

The Low Achieving Managers were Type A style of communicators, using a closed communication style. In fact, they are largely non-communicators.

Thus, High Achieving Managers have respect for their people's ability to perform and expect it (Theory Y). They also use participative management where appropriate, and they communicate openly with their direct reports. High Achieving Managers also have positive attitudes and expectations about their employees' attitudes.

MANAGING AND LEADING AREA 4: MOTIVATION: THE MANAGERS AND THOSE THEY MANAGE

Another topic that many organizational and team decision makers believe is central to performance success is *motivation* (Figure 9.10). As is true with the other areas of management and leadership, having a model as the focus of discussions about effective leadership and management can improve communication. The classic theories of Maslow and Herzberg were a major part of understanding the Achieving Manager's style.

Feedback from the direct reports in the Achieving Manager Project indicated that Low Achieving Managers depended a lot on *safety* and *belonging,* both with themselves and in managing others. Average Achieving Managers focused on *ego* and *safety.* Ego and safety motivation are consistent with the image of a "hard-driving, aggressive, and ambitious" executive. This highly ambitious style of manager has a lot of ego and frequently uses safety as an important part of his or her management of others. High Achieving Managers, while paying some attention to things such as salary, benefits, and equipment, focus heavily on teamwork (belonging), recognition for a job well done (ego), and developing people (self-actualization). Again, the overall consistency of the results from the Achieving Manager Project is remarkable. High Achieving Managers expect their people to perform, they use participative management, communicate openly with their direct reports, and use "higher-level" forms of motivation with themselves and with others.

Two benefits from this view of motivation are important. Astute managers know they cannot motivate someone else. An individual's permanent motivation is internal to that person. The manager can only affect their employees' motivation by identifying what level of motivation is operating within the individual and then providing satisfaction of that motivational need when the person performs.

Second, because motivation is internal to the individual, it is important to pay attention to the motivation of job applicants. As discussed in Chapter 8, there are ways to identify a person's motivation.

MANAGING AND LEADING AREA 5: POWER AND EMPOWERMENT

Managers use power repeatedly. Power is used when a manager has influence in organizational decisions such as hiring and giving direct reports the directions for their performance. The use of power is a centerpiece of leadership and management, and how it is used goes a long way toward defining the style of a leader or manager. As noted above, a manager's power use starts with hiring decisions, goes to defining roles and responsibilities, and involves providing evaluation on how subordinates are performing. At the organization-wide level, power is reflected in defining the organization's strategic direction, setting strategic objectives, and managing how the group does at achieving those objectives. Setting up a system of cascading goals throughout the organization and deciding what standards or key performance indicators (KPIs) should be measured is also an important use of power. As noted previously, leaders tend to get back from others what they are measuring. That is also an important use of power.

The model for assessing power use in the Achieving Manager Project has two parts: (1) power motivation and (2) power sharing. *Power motivation* is a model coming from the famous team of McClelland and Burnam, the former a psychologist who contributed a great deal to our understanding of the achievement drive. Basically, there are three types of power motivation:

1. *Personalized power*: self-serving.
2. *Socialized power*: team oriented.
3. *Affiliative power:* concerned about being liked.

As for power motivation, Low Achieving Managers were seen by their direct reports as having more affiliative power than either personalized or socialized power. Average Achieving Managers focused more on personalized power. While all three categories of managers had each of the power motivations to some degree, the High Achieving Manager's greatest type of power motivation is socialized power. They are motivated most often by seeking to do "what is right for the organization or team."

Power sharing is the second element in the power model. In summary, Low Achieving Managers tend to give away power. Average Achieving Managers keep it for themselves. High Achieving Managers share power with others, which is what collaboration is about. This is not surprising considering their Theory Y values, their attitudes about their direct reports, and their practice of participative management. Remember, however, that the High Achieving Manager is only about 16 percent of the total group of managers.

MANAGING AND LEADING AREA 6: LEADERSHIP STYLE

The sixth and most comprehensive area of leadership and management studied in the Achieving Manager Project was leadership style. The model used to understand leadership style was the Managerial Grid, developed by Mouton and Blake and still very popular today (Figure 9.11), some 40 years after the creation of the Managerial Grid Model.

This model uses two elements of leadership style long discussed in leadership literature, training programs, and discussions. The two elements are *concern for getting the job done* and *concern for people.* Five basic leadership styles result from combining the amount of concern held by leaders and managers regarding the two elements:

1. *9/1 Task master:* powerful concern for the job, little for the people.
2. *1/9 Comforter:* powerful concern for the people and their comfort, little for the job.
3. *1/1 Regulator:* little concern for either the job being done or the people involved.
4. *5/5 Manipulative:* tries to balance concern for both people and job, not too much of either.
5. *9/9 Developer:* powerful concern for both the job and the people.

The results of the achieving manager studies were clear and remained consistent, including the leadership style area. The High Achieving Managers had positive expectations about their people, used participative management, communicated openly with people, motivated others toward higher-level needs, and used power mostly for performing to reach the

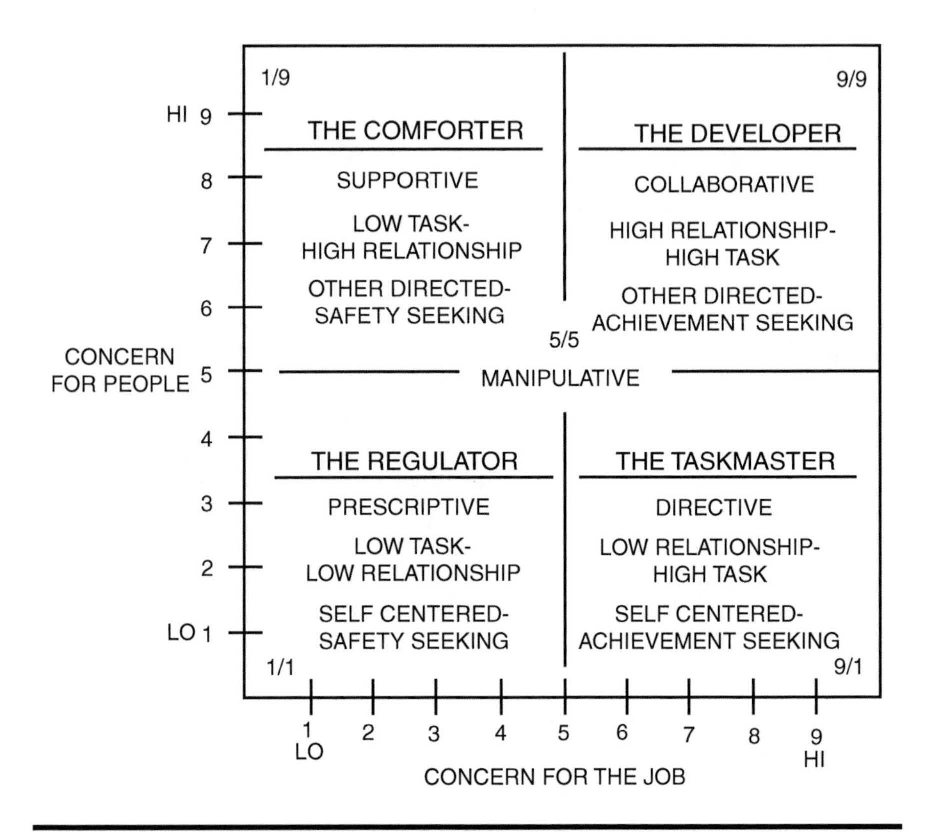

Figure 9.11 Mouton and Blake leadership styles.

goals of the organization or team. It is not surprising, then that the High Achieving Manager also tended to perform as a developer more than the other four styles in the Leadership Grid (see Figure 9.11). The 16 percentage sample of High Achieving Managers out of all managers in the study consistently depended on methods aimed at expecting and getting the best out of their people.

The High Achieving Manager can serve as a very influential model for developing leaders and managers. Experience with a large number of current leaders and managers shows that this influence comes from three sources. First, many people responsible for leading and managing are "in search of a style" on which they can depend. Every day there are new books to read and new leadership programs to attend. The questions many managers ask are: Which style should they believe in and use? What will make them most effective at leading organizations or teams? A brief case study will dramatize this point.

Organization: Ranken Technical College

- Products and services:
  - Technical education
- Markets:
  - Adults
- Structure:
  - Central leadership team of approximately eight people
  - President and vice president with strong personalities
  - Two faculty groups: general studies and technical studies
  - Staff
  - Approximately 3000 students
- Primary issues and problems:
  - Evidence from an organizational culture survey and discussions with some influential leaders in the organization indicated that power was overly centralized with too few decisions made by people who had to implement them
  - Some interpersonal conflict between the two top leaders and others in the organization such that performance was being damaged, in part because of the time and focus spent on unproductive disputes

The leadership group of the college decided to go through a development process in which they learned the High Achieving Manager model. As part of this, they received feedback from at least three direct reports using survey instruments that duplicated the original Achieving Manager Project (Teleometrics). They then set goals and action steps to improve their style to become more like the High Achieving Manager. This included a series of meetings with the direct reports to discuss specific steps in leadership and management improvement. The project lasted about six months.

When the group discussed the (i) participative management and power and (ii) empowerment issues, they came to terms with the problem of an overly centralized decision-making process and decided to push many decisions further down in the organization. The combination of the improved style of the individual leaders or managers and the decentralization of decisions resulted in dramatic improvements in their organizational culture. The college is very successful as an educational institution and is regarded as one of the three best technical colleges in the United States. Now their organizational culture and decision-making processes are also improved.

SUGGESTED ACTION STEPS FOR LEADERS AND MANAGERS

1. Review the status of leadership and management knowledge and skills and style in your organization. Do not depend on just a few opinions; use as broad a source of data as practical.
2. Identify areas where leadership and management is strong according to those being managed and where it is deficient.
3. Use the top group of leaders and managers to begin deciding on actions to improve leadership and management. Discuss how current deficiencies in this area are impacting performance, and how various efforts to improve style might also improve measurable performance.
4. Discuss the various options for improving leadership and management performance with those being managed and led. Recognize that the ultimate test of effectiveness of leaders or managers is the thoughtful opinions of those they lead.

END NOTES

1. Bassi, Laurie J. and Van Buren, Mark E., *State of the Industry Report,* Training and Development, January 1998.
2. Aldag, Ramon J. and Stearns, Timothy M., *Management,* College Division, South-Western Publishing, 1991, p. 500ff.
3. Carol K. Sigelman and David Shaffer, *Life Span Human Development,* Brooks/Cole Publishing Company, 1995, p. 4.
4. Hersey, Paul, Blanchard, Kenneth H., and Natemeyer, Walter E., *Power Perception Profile, A Summary,* copyright 1979, 1988 by Leadership Studies, Inc.
5. For a discussion of the first wave of this study of managers, see the following: Hall, Jay, *Models for Management, The Structure of Competence,* Woodstead Press, 1998.
6. Hall, Jay, *Models for Management, The Structure of Competence,* Woodstead Press, 1998, p. 509.
7. Pace, R. Wayne and Faules, Don F., *Organizational Communication,* Prentice Hall, 1989, p. 121.

Chapter 10

TEAMS: THEIR USES AND IMPACT ON PERFORMANCE

LINKAGE AND OVERVIEW

For many years the fields of Organizational Behavior and Organizational Development have focused a good deal on teams in work organizations. Organizational Behavior has often studied teamwork with particular interest in the performance of teams versus individuals. (1) In general, it is well accepted that teamwork, when done honestly with real issues, increases acceptance of team actions with the team members, improves speed and accuracy in decision making, and improves productivity in some situations. (2) Self-managed teams, those with control over areas such as scheduling their work, hiring into their team, and recognition and rewards, also tend to increase productivity and motivation.

Teams, like any other area of human performance, are not an automatic quick fix. When used the right way and at the right time, they have a fairly predictable impact. To expand what was just listed above, this impact of using teams correctly includes synergy in decision making and improved productivity, often because of anxiety about letting down the team. As discussed above, there is also greater acceptance of team decisions if authority to make the decisions is real. (3) Based on the recognition that these benefits of teamwork are valuable, we have recommended the use of teams in many organizations at various points of performance improvement.

SOME AREAS FOR TEAMWORK

1. The development of the Strategic Plan for the organization or independent team should be done by a leadership group — not by a single individual. The benefits are obvious and are summed

up in the statement that no single individual can make an organization perform successfully, except in a single-person proprietorship. In fact, it takes a collection of interdependent teams or individuals to make a strategy happen. Even in a small organization (e.g., fewer than ten people), the entire group should work as a team to define strategy and accomplish organizational goals. Synergy, which can be briefly defined as increased quality compared to the functioning of individuals working in isolation, is improved with team functioning. Synergy usually occurs in the strategic planning process and in the execution of the Strategic Plan. This greater success in organizational performance is largely due to better quality planning and increased acceptance of the plan from people who had a role in creating it. Motivation to accomplish the strategic objectives is usually greater for those involved in the development of the plan, compared to those handed the plan by their leadership.

2. Management of the actions toward achieving the strategic objectives will improve through a team approach, whether this is the top team under the direction of the top manager or teams throughout the organization charged with achieving "their part" of the objectives. Motivation to keep going on strategic objectives and creativity in deciding on actions are heightened by a good team.
3. Significant decisions about hiring should be made in a team setting. Remember from Chapter 8 that the more information about a candidate one has, the better the decision is apt to be. Input about the hiring decision from all persons significantly affected by the selection is valuable. First, we all look at others from "our own point of view." Having other points of view means we can either confirm or question what we saw in the candidate. Interviewing is affected by the attitudes and background of the interviewer, and balancing and contrasting one interviewer's perspectives with that of others can make the interview data more objective.
4. At the operational level, getting teams involved in setting their shared goals, whether strategic or operational, usually adds to commitment to performance in achieving those goals compared to people doing individual action plans in isolation.

As discussed previously in this book, strategic objectives focus on the longer-term development of the organization or team, with the purpose being to continuously move toward a detailed vision of excellence in performance. However, daily operations require goals that are shorter term — today, this week, or this month. Examples include making an important sale, solving the immediate technology problem, fixing the machine on

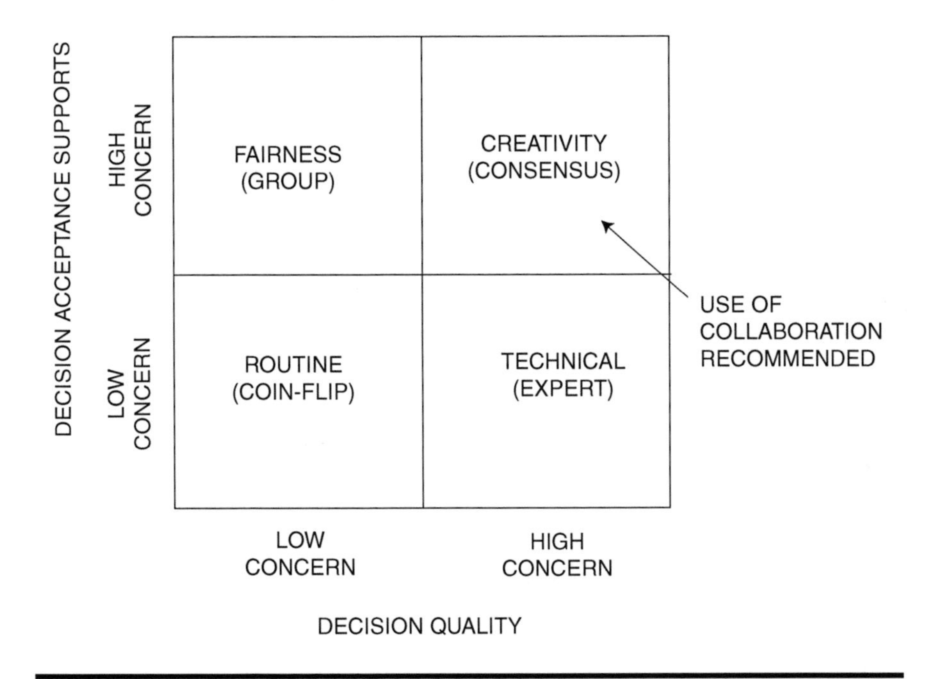

Figure 10.1 Decision process trade-offs.

line 4, and getting the new bank loan. While these immediate goals should serve or at least be consistent with longer-term strategic goals, the pressure of today's activity is real and important in keeping the organization or team functioning. Teamwork on a daily basis, working on activities to achieve immediate operational goals, is also important to successful performance. This chapter discusses the importance and limitations of daily teamwork in supervision and production, marketing and selling, and customer service. Chapter 9 discussed the leadership team.

WHEN TO USE TEAMS AND WHEN NOT TO USE THEM

Chapter 1 discussed the power of fads. This means the tendency of many organizational leaders to initiate popular performance improvement or managerial actions because they have become "the thing to do" for organizational leaders. A specific example of the power of fad is teamwork. Many people automatically jump to the conclusion that teamwork is what is needed in many situations. While the benefits of teamwork are real, as indicated above, there are times to use teams and times when teams are not needed. Briefly review the model in Figure 10.1; it helps clarify guidelines for team use. This model of guidelines for team use was discussed briefly in Chapter 9.

The two dimensions of concern in this model are the extent to which a given decision or project requires decision *quality* and when the issue is decision *acceptance*. As indicated, depending on the project or topic involved, teams can improve quality in decisions as well as acceptance, but teams are not always needed. Here is how it works.

Some decisions fit into the upper left-hand quadrant, where there is high concern for acceptance, but low concern for the decision quality in the sense of the decision's importance to the organization's or team's performance. Again, one example that occurs frequently is decisions about healthcare policies provided to the employees. Many organizations will find a small number of health coverage programs they believe are roughly the same in terms of costs to the organization, administrative responsibility, etc. The decisions about which of the alternatives to make available to the employees in these cases have more to do with the personal lifestyle and attitudes of those employees. Acceptance and fairness are the primary considerations. In this case, a survey of employee desires may be the best approach to making the decision among the options the organization has developed. A team decision is *not* the best approach.

The quadrant in the lower left-hand corner of Figure 10.1 involves the daily routine decisions that involve little concern about the quality of the decision and little concern about its acceptance. The decision must be made but nobody really cares except the person stuck with the responsibility of deciding. Decisions about routine office supplies or what color to paint the restroom are examples clients frequently use. Somebody must decide but, assuming some good judgment on the part of the decision maker, it is their choice. Again, no team is needed.

The lower right-hand quadrant in Figure 10.1 is more complex than the first two quadrants. Here the concern is with the quality of the decision, but its acceptance is much less of an issue. The decision is technical, requiring an expert. Examples here include highly complex IT decisions where only one or two people have the expertise to make the final decision. Another frequent example is the asset investment program of the organization, usually a decision made by the CFO with some direction from the CEO.

Finally, we get to situations where teamwork is needed, which is reflected in the upper right-hand quadrant in Figure 10.1. Here the concern is both with the quality *and* the acceptance of the decision. Examples are developing the strategic plan, setting departmental or team goals, decisions in major projects, and significant hiring decisions. This is where effective teamwork is essential.

People in our business often encounter an issue expressed by hard-driving, success-oriented clients when discussing teamwork. These clients tend to resist productive teamwork and make all organizational decisions through a command-and-control and top-down style. Or, alternatively, they make a weak effort to use teamwork by making team input "advisory," sometimes doing so in the immediate period following a seminar on collaboration they attended.

Work organizations of all types, private or public companies, universities or government agencies, are not democracies. The hard-driving executive might use this obvious truism to resist teamwork, but he befuddles the issue. If organizational leaders want the best thinking as input to decision making and the strongest commitment by people to decisions that have been made, they will find ways to get lots of input to basic decisions and implementation plans. The CEO, president, or department head always has the power and responsibility of vetoing or approving decisions and plans. That is part of their leadership role. Three things are required for the leader to use "advisory" input from her or his teams.

1. Make it clear to the team up front that you can choose to use your "veto" power.
2. Where possible, participate in the team decision-making process but do not over-control the team. This is especially important for a strategic planning team. The leader's active participation in a team decision about a crucial topic almost always means the leader will not need to use an after-decision veto. Team members will pay attention to the leader's well-timed comments during the team's discussions.
3. When a team has made recommendations, and the team leader needs to dramatically change those recommendations or table or dismiss the recommendations, let people know why. That will let them know you took the team's suggestions seriously.

A MODEL OF EFFECTIVE TEAMWORK

Virtually everybody who understands work teams agree on a few basic concepts. First, a team is a group of people working together to achieve a goal or number of goals.

The team can be a temporary project team, such as a group formed to decide on how to meet a technological requirement for a company. This type of team may be a temporary assignment that is disbanded after the job is done — or more precisely, after "the goal is met."

There are many different types of teams: permanent teams such as marketing, production, or sales teams; project teams with a specific assignment and usually a limited existence and discussed above; and top leadership teams charged with providing leadership. These are the most common examples. A less common team is the "self-managed" or autonomous team,

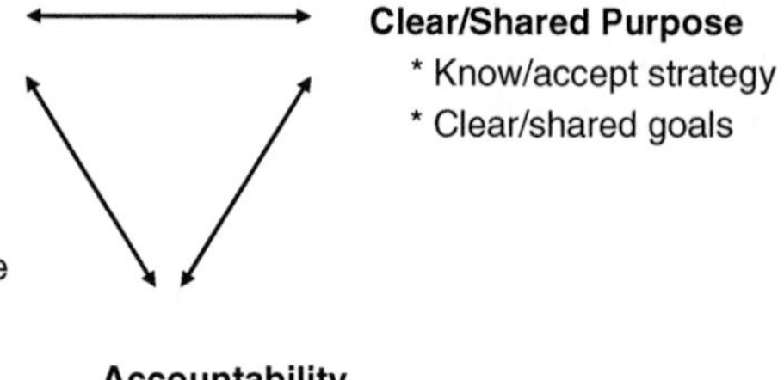

Figure 10.2 Model of Effective Teamwork.

used where responsibility for the entire product or service can be vested in a single group.

Experience shows that there are certain requirements for effective teamwork, regardless of the type of team it is. This is a specific example of a Model of Effective Performance discussed in Chapter 3 and is illustrated in Figure 10.2. (4)

This model is logically obvious, but a brief discussion will help clarify it. To be effective, a team must know what it is seeking to accomplish. Remember that clear goals, strategic objectives, standards, and key performance indicators (KPIs) set the basis for performance, for teams as well as for entire organizations. Without this clear and shared purpose, no one knows specifically what performance is or how the team is currently doing. Therefore, there is no rational basis for performance improvement. As with the entire organization, the team's strategy is in large part the products and services the team provides, and who makes use of these products and services (internal or external customers).

While a clear and shared purpose, the beginning point for team building, is important, the team also must be effective and efficient at decision making. That means the team must make good decisions about its purposes and what actions to take to perform well in accomplishing its goals. This requires three things: (1) open communication and trust, (2) having a process for team decision making that works well for the team, and (3) knowing the content or subject area in which the team is making decisions.

The third component of effective teamwork, *accountability,* is an element frequently missing in teams. The absence of accountability is caused by the same factors that result in the absence of goal setting. People often fear that if they accept accountability for doing something, accomplishing a goal or helping the team perform better, then they will be punished if they fail. So they choose to avoid accountability where possible. The cause for this sad condition is not primarily weaknesses on the part of employees; it is ineffective leadership style. We have discussed this in several previous chapters, and it will be one focus for Part III of this book.

These three components of effective teamwork reinforce each other. If the team has clear purpose, then discussing what to do to achieve those goals or strategy has focus. In addition, defining purpose is part of what teams need to decide. If the team has been assigned responsibility, the team needs to work at clarifying that responsibility. So, clear purpose helps with team decision making, and team decision making must include the following: clarification of purpose and goals and individual responsibility for actions. It is troubling to see how many teams spend a good deal of time in discussions without asking the following questions: What are we trying to do here? Who will take the responsibility for accomplishing that action?

Honesty in team discussions about purpose or the actions needed to achieve the goals is frequently so strained that the benefits of synergy do not occur, and teamwork is ineffective. In this case, the people on the team do not accept or "buy into" the decisions that do occur. The most common example of this is a team dominated by a strong personality or a person in the position of greatest power, such as an owner or CEO of the organization. Leadership has a lot to do with how well the team functions. If leadership sincerely encourages openness in communication, it is apt to happen. If leaders are threatening, then communication will stop and commitment to the goals will be limited. (5)

Clear purpose and goals also impact accountability. No one can be held accountable if the goals or purpose are not clearly defined and accepted. But even if team goals are clear, identifying who is accountable for achieving each goal is essential — and yet frequently does not happen. How many times do team members go away from meetings asking, "Who did we say is responsible for that?"

The relationship between accountability and team decision making clarified by this model of effective teamwork is as follows: the team decides who is responsible for doing what in terms of specific goals. In addition, every team member is responsible for making sure that all decision making is effective and complete. Team members must be willing to raise issues about honesty in the group and the effectiveness of team decision making. In short, they are accountable not only for their part of the team goals, but they are also accountable for helping the team, including themselves, function effectively as a team.

Because a team is a group of people functioning together to accomplish a shared goal or goals, cooperation between team members is needed in many situations. In fact, it is needed any time two or more people are sharing work or have different but related parts of the work process. Also, inter-team cooperation and coordination is frequently needed and yet difficult to accomplish at times. If Team A performs Steps 1 and 4, and Team B performs Steps 2 and 3 in the work process, then inter-team cooperation is needed. Despite the obvious need for inter-team coordination in these situations, it is often deficient; and the deficiency can be caused by factors as specific as physical separation or personality conflicts. (6)

Inter-team coordination and cooperation can be accomplished essentially the same way as teamwork within one group. The key is to develop or clarify common goals between the two groups. Following that, the groups need to learn to communicate with one another to work at actions for accomplishing their common goals. Ironically, sometimes the stronger the teamwork within an individual team, the less willing they are to cooperate with other teams. Still, whatever the reasons for lack of cooperation or outright hostility between two or more teams, actions can be taken to improve their working together. The ACF case study below will help clarify this point.

It is probably obvious that team cooperation is essential in a leadership team developing organizational strategy or reviewing the success of the strategy. It is also needed in project teams and other temporary groups. Effective teamwork is perhaps most critical in the permanent teams or departments in an organization, such as those engaged in marketing, supervision and production, or selling. Because of the ongoing nature of the permanent team members working together, effective teamwork is highly beneficial to performance, and poor teamwork is disastrous. What causes so many teams to be so dysfunctional? Because human beings do appear to have a natural tendency to form groups, why are so many ineffective at achieving goals? (6)

There are multiple reasons for ineffective teamwork. Sometimes the team does not know what it is doing that is dysfunctional or how to fix it. The Model of Effective Teamwork can help there. Sometimes team members do not have the subject matter expertise to make good decisions, so they need either to be educated or to bring in subject matter experts. Sometimes they honestly disagree on basic goals or purpose, in which case they need to develop a strategy. Finally, there is often a personality conflict going on between team members that makes openness in communication or working together on goal achievement more difficult.

Highly assertive personalities often dominate or suppress communication in team meetings, particularly if the power of personality is reinforced by positional power. But two highly assertive personalities can be even more of a team problem, as the two dominant forces struggle over control and power within the group. On the reverse side, if all the members of the team lack assertiveness, the group can be so reserved as to be lethargic. It is possible to attain the right balance of personalities in a working group

— but it is not easy. The following case study will help clarify and extend some of these points regarding teamwork.

Organization: ACF (American Car Foundry)

- Products and services:
  - Manufactures and leases or sells railroad cars
- Markets:
  - Companies that depend on rail to transport their products
- Structure:
  - A single owner who sets major strategy but does not directly manage the company
  - Corporate office separate from manufacturing plant
  - Plant in West Virginia was unionized
- Primary issues and problems:
  - Corporate staff had evidence that communication between plant management and union leadership was especially poor, and hostility between the two groups was impacting performance in negative ways
- Actions:
  - Find a way to build at least temporary cooperation between union and plant leadership on some critical issues

This is a classic example of inter-team conflict. The conflict had a number of causes. It was structurally caused, meaning that the two groups had very different goals and separate legal existence. Second, there were major personality conflicts between leadership at the top levels of the plant and the local union.

Hostility was at an all-time high. The following practical issues existed and little progress was being made in resolving them:

1. There was a major *safety issue* that the groups needed to solve but were having trouble solving.
2. The backlog in labor grievances was unusually high and growing.
3. There was major disagreement between plant management and union leadership over "contracting out." This is the process of hiring external laborers to perform work in the plant deemed to be outside the capability or capacity of full-time employees.
4. Finally, there were hints of concern over quality of the product coming from the plant. Both groups realized that the quality of their products was related to the ability to compete, keep customers happy, and keep the plant operational.

Our intervention in the plant problem identified above involved collecting an efficient survey on how each group saw communication, decisions, and other aspects of the organizational culture. The data was presented on how each group saw the problems and opportunities for improvement, and on how the two groups together saw these issues (combined data). To the surprise of both labor and management, the two groups saw the issues and opportunities in virtually the same way. That set the basis for the possibility of inter-team cooperation. The paper survey data was further reinforced by a series of interviews with the top leadership from both management and labor.

The second step in the intervention was to have each group separately define the four major issues clarified by the survey (safety, contracting out, grievances, and product quality). Both groups identified product quality as a major issue of concern because they believed that the plant had to remain competitive. Each team, labor and management meeting separately, made recommendations about actions they suggested to solve or make progress on each of these four organizational problems.

The final step was to get both labor and management leadership together and discuss the goals, actions, and problem definitions that had been developed separately. The agreement, developed by the separate groups in Step 2 above, on the definition of the problems and the desirable actions was striking. In Step 3 discussed here, the *combined teams* of labor and management then were assigned the goals of taking the input from both labor and management and creating a common action plan. One team worked on *safety,* one team on "*contracting out,*" and one team each on the other two issues. Each of the four "merged" teams developed action steps despite their suspicions regarding each other. Within one year of this intervention, the safety problem had been solved, major progress had been made in reducing the occurrence and backlog of grievances, and some progress had been made in developing a mutually acceptable process for contracting with external labor. Only a small amount of progress was made on the product quality problem, partly because it was the most complicated problem these merged teams faced.

The development of common goals between the two antagonistic groups helped them make progress on immediate issues. Personality conflicts did play a role in inter-team problem solving — as they usually do. These issues are always present, and can only be softened by keeping those in conflict focused on the common goals and purpose. The phrase that works best is this: "I do not have to like you but we are in this together, and our mutual benefits dictate that we need to work together." There was a common purpose in having a safe plant, offering a reasonable amount of extra work to permanent workers, and being involved in fewer grievances. Staying competitive was also essential to both labor and management.

Sometimes, personality clashes or other conflicts such as a basic disagreement on purpose are so strong that efforts to build teamwork do not succeed. This is when reconstituting the team may be required. However, as often as not, attempts at building an effective team suffer from lack of persistence on the part of the leadership of the teams. Team leaders have "accountability" for the development of the team's commitment to their common goals and being effective at decision making. Because of their position, team leaders have the potential for greater impact on the team, for good or bad. The single most important factor in how well the group functions is the attitude of the leader toward the team.

MARKETING, SALES, AND CUSTOMER SERVICE TEAMS

As discussed previously in this chapter, teamwork between permanent department teams is frequently inadequate, thereby creating major problems for the organization. Marketing, sales, and customer service teams are constituted in many different ways and have different responsibilities in different organizations. But a few basic issues are very common. First, sometimes selling is primarily order taking, as is usually the case in E-commerce. But even then, effective marketing, selling, and customer service are needed for customer satisfaction. Second, even when marketing, selling, and service are in the same organizational unit, as is true in the next case study on the advertising company, a clear definition of group or individual accountability can be insufficient. Third, when they are separated organizationally, inter-group teamwork often becomes a central problem. Both organizational arrangements present challenges to effective performance that must be confronted.

To clarify a frequently confusing situation, assume that basic marketing involves product and market planning and research, packaging, promotions, price considerations, and involvement in decisions about which market "segments" to approach. One corporate entity listed the following areas as central to the performance of marketing associates. (7)

- Business acumen
- Customer focus
- Domain expertise
- Entrepreneurial
- Teamwork
- Results focused
- Communication
- Leadership

When the sales function is a separate unit from the marketing group, teamwork and communication has added significance because of inter-team coordination requirements. Selling is closing the deal and getting the customer to buy the product or service. Collaborative support from marketing is essential for sales to be successful. And sales, where it is more than order taking, is where the customers' decisions determine whether performance requirements are met. The two groups need to develop a common purpose, make effective and efficient decisions jointly, and clearly define accountability for each of their units.

When one team shares the marketing and sales efforts, decisions about roles and responsibilities often become a central issue. The following case clarifies this point.

Organization: A 20-Person Advertising Company

- Products and services:
  - Design, development, and placement of advertising
  - Consulting on advertising strategy for customers
- Markets:
  - Chain retail stores
  - Small businesses
- Structure:
  - Managed by the two owners
  - A team of managers who both sold and developed the advertising; group headed by a vice president (not one of the owners)
- Primary issues and problems:
  - The team of advertising people needed to learn to sell, which was one of their responsibilities. Most of them had to make substantial sales as well as engage in marketing functions. This high level of performance was necessary if they were to reach the required corporate income.
  - There were some major issues with the relationship between the vice president and the other managers.

Some people in the group did not want to sell and had little interest in developing their skills at selling. This was true although they needed to accept their selling role to succeed financially. The way the group was structured combined selling and marketing, but accountability was unclear. As a whole, they were a very creative group. However, for at least some of them, they had not translated their creativity into effective problem solving or selling for customers. There was a good deal of resentment and interpersonal conflict over responsibilities for selling versus creating and placing the advertising.

The vice president had not been able to work through the internal team differences over roles and responsibilities, and the fit of individual identities to selling, servicing, and creation of the advertising. Through sales training, the group made progress in learning to sell, although progress varied a lot from one individual to the next. During the training, there were team discussions about selling as related to advertising, and some progress was made on the accountability for these different roles and responsibilities. However, the progress was hampered by distrust of the vice president. The owners, who were getting toward the end of their careers, were distraught over how to fix the issue. Some months after the team completed the sales training, the vice president left the organization and one of the owners retired. Progress had been made on identification of the issue and some resolution of the relationship between the two functions of sales and marketing. In the end, more progress required a change in personnel and leadership.

The vague combination of the dual function of marketing and selling creates many team issues in many groups. That is complicated by the organizational structure dilemma. If the selling and marketing functions are together, the problem becomes who does what. Keep the functions in separate units and the issue becomes inter-group collaboration. In either case, the basic solutions have to come from the teams themselves. They have to take accountability for building common purpose and making decisions about how best to accomplish that shared purpose.

Those planning marketing, selling, and supportive services have additional challenges in today's high technology world. Technology produces situations where the marketing, selling, and servicing mix becomes even more complicated than usual. If the primary contact with the customers is through telemarketing, then those employees "taking the orders" may not understand that they are either seeking to "close the sale" or to deliver service after the sale is complete. Sometimes, telemarketers do both, at times creating functional confusion in their minds and in the minds of the customers. On the other hand, technology offers opportunities for marketing research because of the "reach" of the Internet or related technical systems. Telemarketers can gather at least preliminary data on customer demographics and needs.

These opportunities sometimes lead to the tendency to conduct marketing, selling, and servicing without sufficient planning and clarification to the relationship between these connected functions. When done poorly, this bundling of functions can increase the time cost for the customers, thus creating problems for the telemarketing staff. The major point here is that technology has often confused the traditional functions of marketing, selling, and servicing. Organizational or team leadership is well advised to clarify the ways in which these functions are carried out responsibly.

Selling is identifying customer needs and wants, and placing the features and benefits of the product or service next to those customers' needs as a way of building value for the customers. (8) This value building is important even in the impersonal world of selling through Web sites. Servicing is maintaining or building the value of the product or service through delivery, clarifying product use, repair and return policy, and maintaining collaborative relationships with the customer.

The two are connected: selling depends on customer service to maintain or build value with the customer; and good customer service builds the possibility of additional selling to repeat customers or to people they refer. Selling and service are closely related but they are not the same. The goal of selling is to generate new or additional business. Service is about receiving quality in what has been purchased. Customer service benefits when providers of that service understand both the connection between selling and service, and the differences. Employees vary in their motivation, attitudes, and abilities in selling or servicing, and they should be selected to perform the function for which they are best suited. And customers should not be sold more during product or service delivery, lest they begin to feel manipulated and resentful in the process of being sold.

All work organizations do marketing, selling, and servicing to some degree. Government agencies decide who their clients are, provide information about what they do, and are usually interested in providing services so that their residents are content. For-profit corporations make similar decisions, whether it is a sophisticated marketing, selling, and servicing structure such as that at Anheuser-Busch, Inc., or a CEO planning a marketing strategy with limited data in a four-person company. In any case, organizations need to be clear about who is accountable for marketing, selling, and service, and how they are to be done.

PRODUCTION TEAMS

Teamwork in production teams should focus on the basic function, which is creating the product or service. Products are tangible, while services that we sell or provide are intangible. However, both must be created, modified, or acquired for sale and delivery to the customers or clients. The key is to meet the needs and wants of the defined markets. This is no different for most public or not-for-profit organizations than it is for the for-profit corporations. Increasingly, both for-profit corporations and public or community agencies are constrained by budgets and are evaluated in terms of the outcomes they achieve for the money they spend. For example, cities are concerned with attracting people to their recreational centers and are competing with local golf courses, swimming pools, or other recreational offerings. Churches compete for congregations and provide those services that they think meet their congregation's spiritual and temporal needs.

So, production is important in virtually any organizational setting. People working together to enhance effective production is basic to organizational performance. A review of a case study discussed in a different context in Chapters 5 and 6 will clarify these points and provide lessons for good teamwork in production.

Company: Tone's Brothers

- Products and services:
  - Processed spices
- Markets:
  - Retail chains
  - Food manufacturers
  - Cafeteria management companies
- Structure:
  - Plant of approximately 1000 people
  - About 200 managers, sales, and customer service personnel
  - Factory workers were unionized
- Primary issues:
  - One of the company's plants in an eastern state was being closed. A production operation not previously known in Des Moines was being transferred from the East. The Des Moines plant needed to build production skills for a new packaging line in its local employee base.

The Human Resources department took responsibility to immediately develop training for current and new employees in the operations of the production machine to be transferred. A number of significant factors existed:

1. The product to be packaged by the new line was important to a number of Tone's customers. The purpose was to minimize interruptions in setting up production in the Des Moines plant.
2. The Human Resources team had at least one trainer who had production experience and could with some effort begin to understand the engineering instructions about how the machine operated. This trainer with the experience in production was willing to do what it took to develop the training.
3. The following areas of accountability were determined as a way of developing production training. The Human Resources director would clarify policy on issues such as making worker time available for classroom training on the operation. The trainer with operations

experience would handle translating the engineering description of operations into everyday language. The training designer would focus on writing the training and then with training the trainers.

A team planning meeting occurred early, with decisions made on how to proceed. The goal of developing the training and the associated deadlines was clarified for everyone. Accountability for roles and responsibilities was determined, and training decisions were made, including the mix of classroom versus training on the actual machine. The project team began individual activities with all the pieces of the team plan in place.

After designing the training, there was a train-the-trainer session. One of the decisions was to videotape the machine operations in the eastern state before the plant closed. The decision was made to connect the training material to the videotape, thus adding to the reality of the explanations of machine operations. The trainers being trained had both machine operations experience and some skills and experience in conducting training.

While the pressures of the deadlines created some stress in the project, the team members worked well together. The actual training rolled out on time, and production began very quickly. The training was evaluated by participants and received good ratings for clarity in explanations. The best evaluation of the benefits of this team effort, however, came from the impact on manufacturing the product. The newly trained operators worked the new production line so fast that they ran out of raw materials for the product within a few weeks. The bad news, of course, was that this forced a temporary shutdown. The good news was that the training worked extremely well and produced more product than anyone thought possible.

One other lesson emerged from this project. The Human Resources team had forgotten that it needed to communicate with another group, the Logistics team, about the schedule for training they were conducting. Logistics, therefore, did not know that more production materials were going to be needed quickly.

After a period of operations, Human Resources and the production team decided some additional development was needed. The machine had a number of trouble spots in the packing process, places where problems were most apt to occur. The production team needed to learn techniques for identifying and solving problems as a group to enhance output performance even more. The problem-solving techniques were developed, taught, and used.

CONCLUSION

The use of teams and team-based structures in organizations varies a great deal with the philosophy of the leadership and the nature of the business or the organization. Unfortunately, another factor in team use is whether teams are currently popular in the literature. Current fads in performance improvement over recent years have included teams for customer service, team selling, quality circles, self-managed production teams, and project teams from matrix organizational structures. Despite the ups and downs in popularity, teams will always exist in organizations. The key is to make sure the organizations are using teams at the right time for the right things (see Model for Teamwork and Decision Process). Finally, where teams exist, it is important to make sure they work well (see Model of Effective Teamwork).

SUGGESTED ACTION STEPS FOR ORGANIZATIONAL OR TEAM DECISION MAKERS

1. Identify your intact teams, those that are relatively permanent. Prepare a simple questionnaire using the three components of the Model for Effective Teamwork. Ask team members to evaluate their team performance in these areas. Make sure the survey asks for specific examples and illustrations, plus suggestions for improvement. Make this a confidential survey.
2. Take the results from the survey and begin to work on the areas of deficiency and suggested improvements.
3. Identify the marketing, selling, and servicing functions in your organization. Are they complete, or is some function missing? Is accountability clearly identified?
4. If the functions are separated into different teams, or units or departments, clarify what issues of inter-team functioning may be occurring. Set up a process for finding common goals and effective and efficient inter-team work.
5. Use the Teamwork and Decision Process Model to see if teams are being used at the right time in your organization.

END NOTES

1. A reasonably good survey appears in the early organizational behavior textbook; Hampton, David R., Summer, Charles E., and Webber, Ross A., *Organizational Behavior and the Practice of Management,* Scott, Foresman and Company, 1982, Chap. 5.
2. Aldag, Ramon J. and Stearns, Timothy M., *Management,* College Division, South-Western Publishing, 1991, p. 560ff.
3. Schermerhorn, Jr., John R., Hunt, James G., and Osborn, Richard N., *Managing Organizational Behavior,* John Wiley & Sons, 1991, pp. 219–220.
4. The development of much of this model was influenced by a former colleague, Tony Montebello, and his recent book, *Work Teams That Work,* Best Sellers Publishing, 1994, especially Chapters 3 through 8.

5. Cummings, Thomas G. and Worley, Christopher G., *Organizational Development and Change,* South-Western College Publishing, 1993, p. 217.
6. Pace, R. Wayne and Faules, Don F., *Organizational Communication,* Prentice Hall, 1989, p. 210.
7. Internal document prepared by Jo Hellinger and Dick Goodman, Ralston Purina, 1998. Published with their approval.
8. For an insightful discussion of "communicating value" versus adding value in selling, see the following: Rackham, Neil and De Vincentis, John, *Rethinking the Sales Force,* McGraw-Hill, 1999.

Chapter 11

PERFORMANCE MANAGEMENT: GOALS, FEEDBACK, AND PERFORMANCE

A BRIEF REVIEW RELATED TO THIS TOPIC

A number of major topics discussed previously are centrally related to this topic of *performance management:*

1. Organizations maximize their opportunity to perform successfully when they have developed a strategic plan that outlines a detailed vision of success, including an assessment of their current situation and defined yearly objectives for strategic achievement.
2. Leadership of the organization must manage communication of this strategic plan and the strategic objectives to be worked on currently so that these are understood throughout the organization.
3. Leadership must manage the highest level, or executive team, so that there is buy-in, and active work and support throughout the organization.
4. The organization should have a defined hiring process that targets specific characteristics of the people the organization needs in terms of at least the following areas: the most critical technical knowledge and skills; a strong motivation to learn and grow as the organization changes (which they are all doing in today's fast-paced world); and the values and behavioral style that fit the organizational culture that is most compatible with the strategy of the organization.

Spreading the understanding of the strategy throughout the organization is best done by focusing on goals to be achieved, usually called *strategic objectives*. The communication of primary elements of the *strategic vision* should provide a background and rationale for the strategic objectives, answering the following question: "Why is this objective important?" The assessment of the current situation, as discussed in Chapter 6, provides a basis for knowing what to work on operationally in areas like today's technology or marketing systems.

THE ROLE OF DEPARTMENTS, TEAMS, AND INDIVIDUALS

Leadership in any type of organization provides the strategic direction; that is a critical part of the leadership's job. But most of the work, even in smaller organizations, gets done by people throughout the organization: department managers, team leaders, and employees. Departments or teams of people work together to get the job done assigned to their group. They do this best when they are operating as effective teams, which we described in the previous chapter.

But ultimately, organizations and teams consist of individuals whose performance is what makes the organizational a success or failure. The best strategy will fail without individuals doing the work necessary to make that strategy happen.

We return now to the expanded version of a very powerful quote we used previously in discussing goals:

> "In the late 1960s, Edwin Locke proposed that intentions to work toward a goal are a major source of work motivation. ... That is goals tell an employee what needs to be done and how much effort will need to be expended. ... The evidence strongly supports the value of goals. More to the point, we can say that specific goals increase performance; that difficult goals, when accepted, result in higher performance than do easy goals; and that feedback leads to higher performance than nonfeedback. ..." (1)

So, organizational leadership establishes and communicates the strategy, including organizational objectives (organizational goals for the future). The achievement of the objectives usually gets divided into department responsibilities or responsibilities of teams comprised of individuals from different departments. It is usually desirable to have a number of people working on a strategic objective, although that may be less feasible in smaller organizations having only a few people. But getting strategic objectives accomplished usually happens more often when a number of people are working together on them and sharing ideas and motivational push.

Keep in mind that departments or permanent teams can have strategic objectives but they will also have operational objectives that they work on on a daily basis. These are usually not direct spin-offs from the strategy, although they have to do with the organization's success. As a general statement, strategic objectives have to do with increased success and related growth of the organization. Operational goals or objectives are about success today and they focus on how we get things done.

An actual case example will help clarify this point about strategic versus operational goals.

Call Center at Scottrade Company

(See index for previous Scottrade discussions.)

- Products and services:
  - Answers technical questions for online customers
  - Travel to cover branches during the absence of a branch manager or broker
- External and internal customers:
  - External customers who call and need technical support for their trading needs over the phone or by e-mail
  - Branches that need additional coverage due to an absence of a key employee
  - Other internal departments use the call center as a training ground for people learning the business
- Structure:
  - Three top managers for the call center
  - Approximately 80 employees divided into teams, with each team headed by a team lead
- Problems and issues:
  - Company had a strategic objective aimed at continued improvement of its award-winning customer service

The top manager of the call center served as the chairperson of the firm-wide strategic objective. As a result, a number of actions were taken within the call center to help continued improvement of customer service as part of the strategic objective. In the meantime, the call center continued to work at achieving its operational goals, including responding quickly to calls from external clients, minimizing the number of dropped calls, answering all e-mails within 24 hours, and covering branches where personnel were absent. They also identified and received technical training for their center personnel, and hired new employees to meet their growing workload.

AT THE INDIVIDUAL LEVEL

Where possible, the strategic objectives for the organization should be communicated from top leadership to the department managers regarding what the strategic leadership team wants accomplished. Of course, it is possible that department managers are part of the strategic team, but every department needs to know its role in the accomplishment of the strategic plan. Department managers and organizational leadership can and should agree on the role of each department in achieving strategic objectives.

As the department manager communicates the department's strategic role to department members, this is an ideal time to also identify departmental operational goals. As you may remember from an earlier section, department goals have to do with the daily operations of the department and are not necessarily directly connected to the strategic plan.

Where possible, department managers should make use of the principle of building employee buy-in to departmental goals by having them involved in department goal setting. The manager is still the "boss" and makes final decisions about key goals and issues. But where the employees are experienced in the department's work and motivated to perform at an exceptional level, getting them involved in department goal setting is a wise choice. Not only do department leaders usually get greater support for department goals, at least those that affect their jobs, but the leaders using department goal setting usually get ideas from bright employees that they had not thought of on their own. This is a practical example of synergy.

So at this point in the process, the strategic plan has been developed — or revised if a strategic plan already exists. The strategic goals have been communicated to each department, including agreement on each department's role in the strategy. Departmental operational goals have also been agreed to, with appropriate involvement of departmental employees. Now the stage is set for *performance management at the individual level.*

A MODEL EFFECTIVE PERFORMANCE MANAGEMENT PROGRAM

Performance management programs, sometimes called performance appraisals or performance reviews, have been the target of much criticism over the years. William Demming, an internationally famous personage and a primary force in the introduction of quality programs to Japan, was negative about the use of performance appraisals. (2) Despite our respect for Demming's intellect and ability to help us understand the role of quality in a successful organization, he made the classic mistake of criticizing performance management programs (PMPs). The problem with some PMPs is that they are poorly designed and poorly implemented. Done correctly, they are an important part of maintaining high levels of performance. PMPs can be a major benefit in developing performance at the individual level. And after all, as we have discussed, organizations and departments ultimately are comprised of individuals.

So why do so many performance management systems appear unproductive, leading to calls for their elimination or to managers treating them as a nuisance? As we see from the quotation above, we know that performance is enhanced by having specific and challenging goals, getting people to accept those goals, and providing feedback to those responsible for goal achievement. Therefore, well-designed programs should be a significant enhancement for performance. On the other hand, the dismal record for PMPs comes from poor design processes, a lack of training for the program users, and insufficient support by organizational leadership.

1. *Effective PMP design process.* An effective design process for these programs starts with the recognition of three basic principles:
   a. People are more apt to support the program if their input was considered in its design.
   b. The design of the program must include goals as a primary element of what is assessed about performance. However, those goals should be customized to reflect the nature of the business using the program. For example, performance measurement goals will be different for a manufacturing supervisor than for a customer service call center supervisor or employee.
   c. The design process should include identification and incorporation of significant behavioral style characteristics in the assessment tool. The culture and business of a bank interested in cross-selling its customers, for example, will require customer-friendly service behaviors, whereas the behavior required of a credit and collections officer would be different.

   All three of these basic principles are served by having a representative group of managers involved in planning the broad layout of the PMP, particularly the assessment tool. By having organizational managers involved in identifying the goals and behaviors to assess, the design process dramatically increases the chance of management support of the PMP. In fact, the team planning the broad layout of the PMP should also include well-respected employees.

2. *Training in major topics about the PMP.* Once the PMP has been designed through initial input from managers and employees of the organization, and the initial design is complete, it is time to train those expected to use the PMP. This training should include at least the following elements:
   a. A discussion of the basic layout of the assessment document (goal setting and review tool).
   b. A discussion of the PMP administration process: distribution of the assessment documents, dates for meetings between the managers and their direct report employees (hiring anniversary, end of fiscal year, etc.). These are two-person discussions between the manager and his or her direct reports.
   c. A lengthy discussion of the role of managers and employees in face-to-face performance discussions. This includes the manager's role of coaching, counseling and mentoring.

   The meeting between a manager and her or his direct report employee has potential for substantial fear, surprisingly for both the manager and the employee. Employees usually worry about being criticized by their manager or even being told their job is in jeopardy. Managers, despite having most of the power in PMP discussions, worry about managing the discussion poorly. Most of the stress in these performance discussions, however, comes from a basic psychological need we all share. That is the need to feel good about ourselves, to protect our self-esteem. Protecting our self-esteem requires defending ourselves against criticism. But there is a strong possibility that in a performance discussion, the employees will hear some comments from their manager about their own performance deficiencies. And the employee who feels attacked may attack back and blame the manager for any deficiencies the manager has laid on the employee. In this scenario, both parties can become very defensive.

   There are some techniques that should be included in the training for the PMPs that will help minimize defensiveness. First, managers, the persons with the senior position in the two-person performance discussion, should always remember that "rewarded behavior will be repeated." It is important that the person providing the review include identification of performance strengths on the part of the employee being reviewed. Then defensiveness will be less severe than in feedback discussions where the feedback is entirely negative.

   Second, defensiveness is less severe during the formal performance discussions if employees have received feedback throughout the year regarding the good and the not-so-good of how they are performing. That means that during the annual or semi-annual performance review, there will be no major surprises. The employee is not sitting there waiting for the shoe to drop.

Regular feedback from managers to their direct reports during the year is a part of good management, independent of the role of feedback in a PMP performance review. In fact, periodic feedback is an integral part of the performance management program, and the PMG is a critical part of the manager's role. The training can clarify this connection and the importance of regular feedback, coaching, and counseling from the manager to the employee.

A third technique that can minimize defensiveness, for the employee at least, is for managers to ask for feedback from the employee on how well they are managing. Three questions are especially central to this task. The manager should ask for the following information. "What am I doing in working with you that you wish I would do more often or differently? What am I doing that you wish I would stop? What am I not doing in working with you that you wish I would do?" Organizations sometimes choose to make the feedback from the employee to the manager more structured through the use of a questionnaire completed by the employee. Our experience, however, is that a casual discussion in which questions are asked of the employee about how the manager manages them is the most effective approach.

People do tend to get defensive when they are not sure how to perform some role, such as mentoring, being mentored, and being involved in a performance assessment discussion. The training should ease that potential anxiety by helping people figure out what their role is and how they can most successfully perform that role. This means that the training should be made available for all managers at least. Providing training for employees on their role in the PMP and the assessment discussions is also advisable.

3 *Insufficient support from organizational leadership.* This is a common cause for many performance deficiencies and failure in performance improvement. Organizational leadership, as repeatedly stated throughout the previous chapters of this book, is the most critical factor in high-level organizational performance and performance improvement efforts. That is not to belittle the contribution of the employees or lower-level supervisors and managers to organizational success. We are simply saying that all the evidence and our decades of experience indicate that effective leadership dedicated to high levels of performance is absolutely necessary for making high-level performance possible by the rest of the organization.

The leadership of an organization must be concerned about many things. Getting the leadership to take the time to demonstrate active support of the PMP in their organization is a challenge. However, the following points about the importance of PMPs often influence leaders to become significantly active in their support of these programs.

- Organizational performance ultimately comes down to each individual doing his or her part: upper management, department leaders, team leaders, supervisors, and all employees. PMPs are an important part of managing individual performance.
- PMGs can ensure an adequate spread of responsibility to the right individuals for achievement both of strategic objectives and daily operational goals.
- A well-run PMP can play a significant role in highly important decisions about salaries and bonuses, promotions, and succession planning.
- Active support of the PMP by the top leadership is the single most important factor in how seriously others in the organization will take the process.
- The most important show of support for the PMP that the top leader can demonstrate is his own use of it with his direct reports.

ORGANIZATIONS DOING PMPS CORRECTLY

A number of the companies discussed in previous chapters of this book have made good use of performance management programs.

1. Scottrade Inc. uses its PMP (Scottrade refers to it as a Performance Development System, PDS) for salary and bonus determinations and as input to promotion and transfer decisions. It has updated the PDS a number of times, and has customized the "behavior style" aspect of its program to reflect the desirability of different behaviors for different types of jobs: branch managers, Human Resources, executives, and associates.
2. The chiropractic organization discussed in Chapter 2 has just begun the use of the PMP as of the writing of this book. Despite being a relatively small organization (approximately 20 employees), it has found that discussions about current performance and how to improve performance have been useful. It has also helped the two owners avoid their previous tendency to blur areas of responsibility, with the result that employees sometimes received conflicting directions from the ownership.

3. Landshire Company, the sandwich manufacturing organization discussed in Chapter 4, has customized its PMP to reflect the rapid pace of a route delivery firm. It has also connected its program to bonuses and promotions for the route drivers and others throughout the organization.

SUMMARY AND CONCLUSION

This chapter discussed programs throughout organizations referred to variously as performance appraisals, performance management systems, or performance reviews. Whatever they are called, they tend to be unpopular; and some notable persons have suggested that they be eliminated from organizational systems and procedures.

We have argued that the primary problem with performance management programs (PMPs) results from how they are created and implemented. We have suggested that working carefully with three areas of design and installation will minimize issues and allow PMPs to have a significant positive effect on performance.

SUGGESTED ACTION STEPS FOR ORGANIZATIONAL OR TEAM LEADERS

1. If there is a PMP in operation, conduct a confidential survey on what people see as the benefits of that program and areas where it needs improvement. Make the survey confidential, but have the respondents mark whether they are managers or employees. Make changes as necessary.
2. If there is no PMP, follow the guidelines in this chapter to create one.

END NOTES

1. Stephen P. Robbins, *Organizational Behavior,* Prentice Hall, 1998, p. 180.
2. Peter R. Scholtes, *The Leaders Handbook,* McGraw-Hill, 1997.

Chapter 12

RECENT PROGRAMS EMPHASIZING EFFECTIVENESS AND EFFICIENCY

HOW THIS CHAPTER FITS IN

One of the primary goals of this book is to help leaders and managers see and understand a comprehensive approach to evaluating the performance of their organization, department, and team. The second primary goal is to provide those reading this book with a major overview of a number of programs (interventions) that can be used to improve performance. Our perspective about what interventions work and in what situations comes from a combination of numerous decades of assisting organizations with performance improvement plus the use of the appropriate literature in organizational development and related fields.

In referring back to the Organizational Success Model, previously discussed a number of times, one will recall that at the general level, successful organizations have two primary elements well developed: (1) they have a well-designed strategy that they clearly communicate to their employees; and (2) they have effective and efficient operations. Operations include how they manage, sell, deliver their products and services, etc.

When an organization is effective, it is able to get things done with good quality. So, the effective restaurant delivers a good dinner, the effective shoe manufacturer makes shoes that lasts, and the effective CPA firm completes tax returns that pass muster with the IRS. Essentially, the effective organization gets things done that hold up, meet a need, or are of good quality. Holding aside issues of price and time of delivery, their products and services meet customer needs and are in demand.

On-time delivery and costs are efficiency issues. Efficiency means that the organization creates, produces, and then sells and delivers its products at a competitive cost of money, time, and other inputs. Efficiency is defined as a lower level of inputs to outputs. Efficiency focuses on the amount of time and resources (both human and raw materials, etc.) that it takes for the organization to create and have available whatever product or output is important to its success.

Two further refinements on what is meant by efficiency are needed. First, we use the term "products" here to include both what the organization receives income for doing and what the organization needs to create or develop for its own internal uses. An example of the former is anything, tangible or intangible, that generates income for the organization, such as software for Microsoft and college courses for a university. An example of the second type of product would be Microsoft adding a new call center for its clients with technical questions to access at no cost to them. The call center does not directly make money for the company, although it helps satisfy customers. For clarity, it is important for our readers to know that the term "products" used in the numerous cases in this book refers only to products that generate income for the organization. The latter type of products, such as a call center, usually only costs the organization, but someone in the organization believes they are important for the organization to have.

Efficiency means that the real cost of producing a product or service is low. Therefore, the cost of production for the first type of product, which is provided to an external customer, will be reasonable compared to the price of an inefficient organization. Because the time it takes to create or deliver a product or service is part of the cost, the expense for a product or project to be used internally by a company will be paid by the company. Another example of an internal product is creating a new Web site to represent the business. The cost for this must still be paid. Someone has to pay for the time and materials for the Web site, even if it is an output that does not itself generate additional income for the organization. Organizations have many internal systems and programs (the latter type of product) that they believe they need to have for their organizations. Remember that products and services, both those sold to customers or clients and those used internally, have costs that someone must pay. This is true whether the product is tangible (e.g., a CD purchased at the grocery store) or an intangible service (e.g., financial advice from a CPA).

Programs to improve effectiveness and efficiency have been used by organizational leadership for many decades. As competition has continued to increase for U.S. organizations, there has been increased interest in these programs. Many observers of the United States and other developed economies say that the demand for increased speed of product or service creation and delivery will continue to grow. This demand for speedy delivery is usually accompanied by the requirement for competitive pricing of these high-demand products and services. That will probably mean that programs that focus on effectiveness and efficiency, such as those described in the remainder of this chapter, will continue to be popular.

PROJECT MANAGEMENT

Project management (PM) is an overall approach for managing many different programs within an organization. Some of these projects are heavily oriented toward the types of programs and interventions discussed in this book. Many of the programs worked by project management today are about technology or are financially focused. PM has recently become integral to the strategic and operational aspects of many organizations, which is why it is included in this new edition of the book. Basically, PM is related to the focus of organizational development (OD) — and what is discussed throughout these pages — in two distinct but related ways. First, PM makes use of many OD programs and interventions in the management of its projects: hiring and selection of project team members, team building, relating the project to the organization's strategy and organizational culture, and leadership/management, just to name a few uses.

Second, many of the (OD) interventions discussed previously can make use of PM as a way of managing the OD intervention. This most obviously includes a strategic objective such as a major technology addition to the organization. Revisiting a case from Chapter 3 is useful here.

The tool and die company discussed in Chapter 3 involved a strategic planning process with the company leadership and a facilitator. One result was a strategic objective that required the purchase, installation, and implementation of computer-aided drafting (CAD) and computer-aided manufacturing (CAM). A strategic objective such as this can benefit from a thorough use of project management. This was a major undertaking by the tool and die company, and was expensive in terms of both time and money. Using PM fit this company's CAD/CAM purchase.

The "project" in project management is defined as follows: "A temporary endeavor undertaken to create a unique product, service, or result." (1) What follows is a brief discussion of the characteristics of projects in PM taken largely from the American Management Association's *Handbook of Project Management*. This list will further clarify what this type of program can mean to performance improvement effectiveness and efficiency in both technology and other types of interventions. (2)

1. Projects are unique undertakings in the sense that they are distinct from anything else done by the organization. Using the tool and die company as an example, numerous organizations have installed CAD/CAM. The tool and die company was doing this installation for the first time.
2. Projects are composed of interdependent activities. A project to be worked by project management is a "system" consisting of interdependent parts that must be managed for effectiveness and efficiency. In the CAD/CAM example, the tool and die company wanted to get the systems chosen, installed, and operational so that they helped the company in its business strategy. The company wanted to make sure that it chose the right versions of CAD/CAM.
3. The company also wanted to make sure that the systems were chosen and made operational at reasonable costs of money and time.
4. Projects create a quality deliverable. The deliverable in the example here was the new CAD/CAM systems. The quality was based on whether the users of these manufacturing support programs would be satisfied with how they worked and helped the company meet its business goals. Of course, getting the systems operational is the start of good quality. Quality also means few defects in the way something works.
5. Projects involve multiple resources. The purchase-to-operations process of installing the CAD/CAM took time; the involvement of engineers, executives, and financial people at the company; and lots of money.
6. Projects are not synonymous with the products of the project. The product was the actual machinery and software of the CAD/CAM. The project to be managed included the early planning about what the company needed these systems to do, who would make the contacts and negotiate with the providers, and all of the various steps (including training of the system operators once the systems were installed).
7. Projects are driven by the triple constraints of time, resources (human and non-human), and quality (does it work well?).

A project in project management operates through the stages of a project life cycle. A project life cycle is defined as having the following stages:

- *Concept.* This is the planning and layout of requirements for what the project output or product should provide. In the tool and die company example, this started with strategic planning, which led to an objective for the purchase and installation of the CAD and CAM systems. However, this was only the start of project management of the CAD/CAM program.

- *Development.* In the CAD/CAM example used here, this included contacting potential suppliers, getting cost estimates, identifying different providers of the systems and how they compared, deciding on timetables for various steps, and comparing various systems to the requirements identified by the tool and die company.
- *Implementation.* Obviously, this is getting the products (again in this example, CAD/CAM programs) operational so they can be effectively and efficiently used.
- *Termination.* This is the final step in the project life cycle. Years after it had implemented CAD/CAM, the tool and die company is still using it, although some enhancements have been made along the way. What has been completed (and therefore terminated) is all of the activities to do the original, unique purchase and installation.

Many other strategic or operational programs in an organization's development could meet the characteristics of a project in "project management." Therefore, major performance improvement interventions may need structured project management to accomplish the OD project effectively and efficiently. Some possible PM uses connected to organizational development might include new product research and development coming from the Strategic Plan or marketing research developed by a marketing or sales department's leadership. Further examples include Scottrade Inc. using project management extensively in a number of its recent IT endeavors. For that company, this has included working to increase the capacity and stability of the company's trading Web site, and expanding and moving its data center from headquarters to a larger facility.

The following provides a few other examples of how PM might be the program by which OD projects and goals are accomplished.

- Project management would be useful if a company decided to develop and install a major training program where it had limited training in the past and could have been primarily using external suppliers.
- Project management could be the system managing a project where an organization is evaluating whether to enter a major new market, either a geographic or demographic market. New market decisions can have huge implications for an organization, a corporate firm, a university, or even a religious institution considering a new country for charitable works.
- If an organization is planning a major restructuring of the entire structure or a substantial part of it, project management would be useful. (We discuss this further in the final section of this chapter.)

The previous few pages here have discussed how PM might be used by OD in planning and managing various types of OD interventions. There are also many ways in which PM uses OD in *its* programs. PM makes use of OD in many areas of teamwork, which is a key part of many performance improvement interventions. Various management and leadership techniques, usually included under OD, are also key to successful PM. In fact, there are many other OD programs and techniques used in PM, including group theory, interpersonal skills, and communication. (3)

AN OVERVIEW OF THE CONNECTION BETWEEN PROJECT MANAGEMENT AND ORGANIZATIONAL DEVELOPMENT/PERFORMANCE IMPROVEMENT

The two disciplines of *project management* and *organizational development/performance improvement* overlap significantly. However, each also has areas of professional work and process that are outside the other. PM, for example, can have many areas outside the "human/organizational" focus of OD/performance improvement. Information technology improvements and major construction projects are examples.

Many product development projects are way beyond where organizational development would be directly involved. Thus, an organization going through strategic planning might decide on a new product that, because of its complexity, is designed and developed through a PM process. In this example, after the initial stage of strategic planning, OD would probably have little to do with the product development done by PM. On the other hand, the strategic planning process, which is an OD/performance improvement intervention, will often be the cause of programs needing project management.

To use a reverse example, PM might be suffering performance deficiencies because of a lack of teamwork, which is a key to project coordination and collaboration. Or perhaps the PM leadership team would benefit from learning some additional management skills. This training would come from the training section of OD. To re-emphasize a point made previously, OD and PM overlap but the two fields have some areas separate from each other.

TWO MANUFACTURING INTERVENTIONS RELATED TO PERFORMANCE IMPROVEMENT: LEAN MANUFACTURING AND SIX SIGMA

The U.S. economy and other advanced economies, such as those in Western Europe, have become more technical and service oriented and much less manufacturing oriented compared to a few decades ago. Still, manufacturing is an important part of our economy, especially in industries such as housing materials, automobiles, healthcare, and technology products. Our economy needs manufacturing for some of the more complicated products. Therefore, we have been striving to make our remaining manufacturing industries more defect-free and profitable. As a result, we are now working on a number of programs where the goal is products with few if any defects, produced as rapidly as needed by the customer, and at competitive costs.

Keep in mind that products have a physical existence, like the chairs and table at breakfast, the cars we drive, and our cell phones. Services are some type of behavior or knowledge that is valued and paid for by someone. When we see our medical doctor or dentist, we pay for their knowledge and advice, which are services, although we may also pay for medical products such as medicines and bandages. Here we are using the terms "products" and "services" to mean something that is paid for by an external client of the provider, someone outside their organization.

Lean manufacturing focuses on making tangible products (e.g., tools and dies or automobiles) defect-free and cost competitive. It focuses on effectiveness by working to make the production process and the resulting products as defect-free as possible. The following are a few of the techniques used by lean manufacturing primarily to increase production effectiveness:

- *Bringing inventory into the production facility just-in-time.* This includes the raw materials for production of the finished product. "Just-in-time" means just before it should be sent to the customer. This is thought to help keep costs down because having materials or finished product sitting in inventory results in money tied up in the costs of labor and buying and processing those materials. This focus is based on the financial principle of the "time value of money."
- *Maximizing customer satisfaction* through on-time delivery of the product, making it as defect-free as possible, and keeping costs as competitive as possible. Six Sigma has become popular recently as a process for measuring how well the production process is working. Historically, measuring a production process was popularized in the manufacturing world about 20 years ago under the name "statistical process control." It was a spin-off from the work of two quality pioneers, Demming and Juran.

A number of manufacturing firms are combining *lean manufacturing* with *Six Sigma,* a use of data to measure both the effectiveness and efficiency of the production process. The data of Six Sigma focuses on maintaining or achieving high quality, which essentially means defect-free products meeting customer specifications. Obviously, the fewer defects in the product, the lower the costs to the producer and possibly the customer because there is less need for re-running a product to fix defects. All of these product characteristics resulting from programs such as lean manufacturing and Six Sigma are targeting effectiveness and efficiency in the production process. When the programs work, the result is *quality* products.

Services — something of value the customer is willing to pay for that does not have a tangible existence — are sometimes targeted for efforts at quality. If, for example, a CPA firm wanted to maximize the quality of what it provides its customers, it could work to improve how rapidly it generates financial reports. However, this example illustrates how measuring the quality of services is different from measuring the quality of products. Obviously one knows if one's new car is not running or one's camera cell phone does not take photos. In an accounting firm, quality would be measured by making sure there are no mistakes in the financial reports or audits that the CPAs create for their customers, but that assumes the ability to get another CPA to evaluate those reports. A part of the quality of services like financial reports is on-time delivery. In addition, competitive cost of services can be used as a quality measure of services. The major difference between quality measures of products versus services is knowing that the service provides what is needed. That is the essence of effectiveness.

A case from a manufacturing firm will help clarify lean manufacturing and measure through Six Sigma or similar statistics for process control.

Ehrhardt Tool and Machinery Company (see Chapter 3)

- Products and services:
  - Tools and dies for use by other companies in their operations (e.g., nuclear fuel plants, etc.)
  - Machinery made to specifications
- Markets:
  - Original equipment manufacturers, e.g., automotive plants, etc.
  - Nuclear fuel plants
- Structure:
  - Ehrhardt is a wholly owned subsidiary of a company holding eight other manufacturing companies
- Issue:
  - The holding company wanted to build competitive advantage for Ehrhardt by making their product of the highest quality possible

To maximize the competitiveness of its products, Ehrhardt focused early lean manufacturing efforts on improving efficiency in the shop. They added shop carts so that all tool makers would have their portable tools immediately on hand. They also videotaped the tool production process to identify any areas of delay in the process. They worked at making the end-to-end production process more efficient by reducing the *production time* for products and by reducing *waste and defects* in production.

Reorganization can serve as an efficiency and effectiveness intervention if the organizational leadership initiating the restructuring uses these factors as a primary reason for its efforts at reorganization. In fact, improving the ability of the department, team, or organization to get its goals accomplished effectively and efficiently is *the best reason* for restructuring. Restructuring to reduce costs is justified if the organization's survival is in jeopardy. However, any reorganization effort will likely lead to at least a temporary reduction in production, whether in products, services, or customer service. Therefore, restructuring initiatives aimed at cost cutting should be done with awareness of their potential negative impact on effectiveness and efficiency.

Organizational structure, reporting relationships between positions, is what reorganization targets. Those performing restructuring should keep the basic building blocks of organizational structure in mind. Those building blocks of organizational structure include division of labor, departmentalization, chain of command, and span of control (4).

Division of labor is the separation of tasks into separate jobs. It is a key aspect of how the work in the organization gets done. One of our associates had the following experience; it illustrates the importance of division of labor to customer service. He entered the local branch of a national copying company and approached the desk that had a sign indicating that the section specialized in computer documentation. He had the document he wanted copied on a computer disk, and asked the person at the desk if she could convert the document from one word processing format to a different one. Our customer had asked for that. The person from the copying company indicated that the conversion would be no problem. Our associate then indicated that the conversion service was appreciated, and asked the person to please make 20 hardcopies of the converted document. So far, so good.

The person from the copying company looked at our associate and indicated that she could not have the hardcopies made because that was "a different department." Somewhat confused by this response, our associate asked whom he needed speak with to get the document printed after it was converted. The copying representative pointed to another employee sitting at the desk next to hers. The second copying company representative, who had clearly heard the initial discussion, sat waiting for our associate to repeat her request for copies of the document once the technical conversion was complete.

The brief situation above is a problem caused partly by poor customer service and partly by division of labor. But the point is made: reorganization should be done with an eye toward effectiveness and efficiency for the customer, which is a major part of good customer service. Which tasks should be separated and which should be kept together is more than just a question of how someone sees the lines being drawn on an organizational chart. Ultimately, effectiveness means how well does what we do meet the needs of our clients, students, or medical patients and their families.

Departmentalization is the grouping of jobs into organizational units on a more or less permanent basis. The most common basis for departmentalization is by function (e.g., Human Resources, Finance, Engineering, and Production). However, departmentalization can also be done on the basis of grouping by tasks related to the same products or customers. Sometimes, departmentalization is geographic, such as branches in a national bank with locations across the country.

Some organizations have departmentalization for their own convenience rather than efficiency or effectiveness in creating or delivering their product or services. The chiropractic group cited in Chapters 2 and 11 used its strategic planning correctly by departmentalizing its services according to patient convenience. First, the group planned additional offices according to patient convenience in getting to the offices. Second, the process of verifying the patient's health insurance is done as completely as possible on the phone. The goal is to have a minimum of health insurance issues requiring resolution at the beginning of the patient's first visit. By the time the patients are ready to leave after their adjustment by the doctor, they know exactly what their insurance will cover for future services.

This is what the chiropractic organization did to make this happen. It kept the receptionist department involved in a number of the steps of gathering insurance information from the patient. It also had the insurance verification department closely coordinate with the receptionist department on the insurance for any patient. Therefore, each department (receptionist and insurance verification) was aware of the patient's health insurance coverage and any additional costs remaining for the patient. This benefits the organization because it reduces duplication of insurance verification. It helps the patients because they know what their financial obligation will be before they decide on services beyond their initial visit. This is a win–win situation in which the healthcare corporation also receives benefits. The corporation get its cash flow more quickly than if it simply processed billing after the services are delivered. It also has an unusually high patient retention rate.

Chain of command is the line of authority that extends from the top of the organization to the bottom of it. How the reporting relationships are organized and who reports to whom is the central topic. There are really two primary decisions within the broad area of chain of command. One of the decisions has to do with how many levels of authority the organization will have and what they should be named. Thus, first is the president or CEO of the organization. After that, a standard approach is to have an extended executive level including various vice presidents, followed by top managers, middle managers, supervisors, and team leads. Some organizations, particularly large ones, can have all these levels before they ever get to the employee level.

The second primary chain-of-command decision organizations must make is the amount of communication between the levels of the chain of command. A related topic is how different levels communicate and how often higher levels are accessible to the levels below them. Sometimes the top levels of an organization release general policy, such as sales or production goals, but leave the detailed decision making and strategy to the departments below. While this can be seen as a positive form of "empowerment," it should be remembered that empowerment works best when there are guidelines from top organizational levels about expectations regarding client service, costs of operation, and hiring responsibility, among other considerations.

The decisions about chain of command often generate great passion because decisions about who has what authority are allocations of power, prestige, and rewards. Still, as difficult as it might be in chain-of-command decisions, the primary guidelines in making these organizational levels decisions should be effectiveness and efficiency. In summary, the question is how to structure from top to bottom of the organization so as to maximize effective and efficient delivery of all products and services. This is an important guideline, no matter what kind of organization is involved.

Span of control refers to the number of people any manager can manage (provide directions, coach and counsel, do performance reviews, etc.). There have been many studies trying to identify a range of reasonable span of control. In general, organizations frequently have managers and supervisors with a span of control that is dysfunctional.

> "At some point wider spans reduce effectiveness. That is, when the span becomes too large, employees' performance suffers because supervisors no longer have the time to provide the necessary leadership and support." (5)

Span-of-control decisions are sometimes made based on criteria other than the effectiveness identified in the above quote. As often as not, how many people will report to a particular manager, supervisor, or team lead is based on functional criteria, meaning combining all the workers who essentially do the same kind of work. But how well they do that work is usually impacted by how well they are managed and supported. An overly large span of control can dramatically reduce employee effectiveness at getting the job done.

Span of control is one of those areas of structural organization that usually requires some balance between effectiveness and efficiency in determining the span of control size. The following from our work with Scottrade Inc. will clarify the issues here. To review the details of this company's basic products and markets, review Chapters 6, 7, and 11.

The CEO of the company was the founder and remains the primary principal in this rapidly growing private company. Organizational climate surveys confirmed our impressions that he is extremely well regarded by those working in the company. Therefore, everyone wanted to maintain contact with the top manager and be his direct report. As a result, he had more than 20 direct reports. He felt he was out of contact with much of what was going on in his business. Evidence supported that impression.

To correct this situation, some reduction in the number of those reporting directly to him was needed. However, that presented major problems for those who would no longer be his direct reports. The lengthy discussions about reduction in his direct reports included thorough consideration of the following topics:

- Which of our departments with direct impact on client services need to have direct contact with the CEO through their top executives?
- Which departments are most central to the achievement of our strategy? Those departments should be primary candidates for direct reporting to the CEO.
- How do we make it clear that those who are no longer the CEO's direct reports will still have the right to communicate directly with him? He will talk to anyone when it is important.
- What is the best way to use the top group of executives as a central discussion and decision-making team? How should their areas of responsibility relate to that of our Research and Development committee and its obligation to focus on new products and programs?

These were the primary considerations in the discussions of how to restructure the top level of the company. While morale was a consideration and was discussed frequently as the decisions were made about reorganization, the business issues listed above were primary in this effectiveness and efficiency case.

CONCLUSION AND SUMMARY

Several pages in this book discuss goals and the measurement of actions related to those goals (such as key performance indicators, KPIs). Two of the most important KPIs are *effectiveness* and *efficiency*. What should be specifically measured as the effectiveness and efficiency KPIs should vary with the type of organization. Specifically, what are its products and services, markets, and methods of creating and distribution — what it provides of value? But these are the best guidelines for deciding what to measure and manage.

- *Effectiveness*. What are the areas of central importance regarding how well we accomplish our goals: quality of products and services, customer service, hiring, financial ratios such as cost to output, etc.?
- *Efficiency*. What efficiencies are most central to our success: on-time delivery, speed of call center responses to phone calls, advertising cost for our new business leads, etc.?

SUGGESTED STEPS FOR ORGANIZATIONAL OR TEAM LEADERS

1. Answer the questions contained in the two bullets of the final paragraph above.

END NOTES

1. Dinsmore, Paul C. and Cabinis-Brewein, Jeannette, *Handbook of Project Management, second edition,* American Management Association, 2006, p. 2.
2. Dinsmore, Paul C. and Cabinis-Brewein, Jeannette, *Handbook of Project Management, second edition,* American Management Association, 2006; see page 2ff.
3. Dinsmore, Paul C. and Cabinis-Brewein, Jeannette, *Handbook of Project Management, second edition,* American Management Association, 2006; see Chapters 12, 12A, 12B, 13, and 13A, pp. 136–183.
4. Robbins, Stephen P., *Organizational Behavior,* Prentice Hall, 1998; see Chapter 13 for a full discussion of organizational structure components.
5. Robbins, Stephen P., *Organizational Behavior,* Prentice Hall, 1998, p. 483.

Chapter 13

PERFORMANCE IMPROVEMENT: CHANGE, LEARNING, AND DEVELOPING THE INDIVIDUAL

INTRODUCTION

Thus far we have discussed performance improvement at the organizational, team, and individual levels. This has included descriptions of some interventions, or ways of working to develop increased performance. Interventions discussed to this point have been strategic planning and cascading goals, building a learning organization, hiring, developing managers, team development, some more recent programs focused on organizational effectiveness and efficiency, and performance improvement programs. This chapter focuses completely on the role of the individual employee and his or her performance in all of this.

1. Individuals create the strategy, set the objectives, and either do or do not make the strategy happen. The CEO and individuals on the leadership team can create the strategy, support and recognize performance, and deal with reducing performance gaps. But success in achieving high-level performance resides with all the organization's individuals, including those in leadership. Every aware individual in the organization will make a personal decision about

"buy-in" to the Strategic Plan or other performance improvement intervention. This decision about buy-in to the performance improvement interventions will then dictate the individual's attitudes, motivation, and effort about making the strategy work. It is a wise organizational leader who understands the power of individual decision making throughout the organization that he or she leads.

2. While individuals are influenced by department or team goals, roles, and cohesion, individuals in all sections of the organization — marketing, sales and service, and production — will react to the "team-ness" in their own unique ways, for better or worse.
3. Perhaps nowhere is the power of individuality so apparent as in how successful managers are in managing their teams. The Achieving Manager Project, for example, discovered that among 16,000 managers, there were varying degrees of leadership success. High Achieving Managers dealt with their direct reports much differently than other, less successful managers (about 86 percent of all managers). In fact, experience shows that individual managers vary a great deal in how they use the six leadership areas identified in the Achieving Manager project (see Chapter 9). Their style of managing their team also influences their success in getting team performance. However, the individuals within the team still make their own decisions about how hard they will work to perform.

Any individual's performance is an inter-play between organizational characteristics; team dynamics; how well they are managed; and their individual identity, behavior, and goals and motivation.

Do organizations and teams have influence over the performance of individual employees?

Despite the fact that individuals ultimately decide about their own motivation and desire to perform, organizations and teams do influence their decisions about such things in many ways. Organizations hire and promote certain individuals and not others. Managers and supervisors direct people more or less effectively, and they recognize, reward, and sometimes punish individuals. In short, they provide opportunities for learning and performance growth, or they provide conditions that result in performance failure.

A simple model, Identity/Role theory (Figure 13.1), helps clarify this.

Identity is who we are, the same individual personality, motivation, attitudes, and values characteristics discussed under identity in the Model for Growth. Two primary forces impact the way we perform: (1) one's identity, and (2) what one knows how to do and what one believes about the role one is performing. Some of our *work roles* are heavily influenced by the organizations in which they occur (e.g., supervisory roles in a

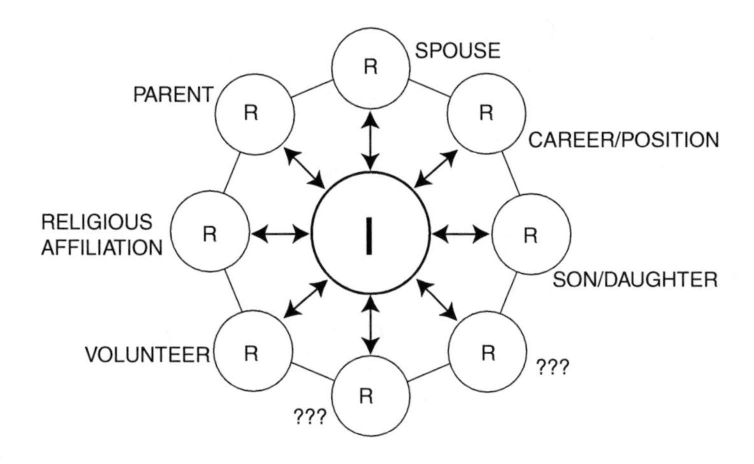

Figure 13.1 The Identity/Role theory.

manufacturing company, route drivers in a delivery setting, etc.). Our *personal roles* are defined more by the culture and socioeconomic levels in which our personal lives exist. Our personal roles are also affected by our lifestyle decisions, those pertaining to our family arrangements, educational achievements, and occupational choices.

All of our roles in life are connected in the sense that how we are doing in one role will probably affect how we do in other roles. This is very true regarding the connection between most of our personal roles and our professional role. Few people succeed for very long at separating their personal and occupational lives. When things are going well personally, it is usually easier to perform well at work, even if work is not everything we would like. If things are going well at work, our personal lives benefit as well. The following simple formula helps communicate this:

A Fulfilling Life = Good Love + Good Work

So things that happen in our personal and work worlds affect one another. As one CEO was heard to say, "When big problems occur in my company, I go home and kick the dog." She might have also said the following: "When things go well at work, I take the family out for a celebration." The complicated interaction between the many roles in life is illustrated by Identity/Role theory.

While the importance of personal versus occupational life varies with the individual, there is little doubt that they are interconnected. Even examples of people who have little personal life and try to "lose themselves in work" support this conclusion. They may well be motivated to substitute work for lack of personal satisfaction. The fact that employers recognize the personal life–professional life connection is demonstrated by the popularity of "employee assistance programs" and the effort by many companies to provide educational programs in personal finances. Problems relating to personal finances are among the major causes of difficulties at work and missing work.

WHAT IS KNOWN ABOUT DEVELOPING INDIVIDUAL PERFORMANCE

The discussion thus far presents a dilemma with regard to performance. First, individuals ultimately decide how much they will put into doing their work, how much they will put into performing. Individuals will also decide how much they will contribute to the special efforts required of performance improvement interventions such as strategic planning and working the strategic objectives, making major efforts to hire the right people, or going through the stress of conducting a performance management meeting with employees.

The second part of this dilemma regarding performance is that the individual's work performance is affected by other parts of his or her personal life — family relationships and the like. Therefore, even the best manager who has an extremely high-performing individual or team, may suddenly find that some or all of his performers have become less productive, and the reasons seem to have nothing to do with work conditions. But just as discussed about hiring in Chapter 8, getting successful performance from an individual is a matter of probability, not certainty. What leaders and managers can do is increase the odds that their employees will perform by doing everything to maximize those odds. The techniques for maximizing the odds of high performance are included in the best publications on organizational development and organizational behavior, and management writings from people who have led and managed successfully or helped those who have.

The ability to maximize the odds of a high level of performance by those one manages and leads is improved by knowing about how to work for individual development.

Two conclusions about developing individual performance flow from the above discussions:

1. Any performance improvement program should make efforts to develop the individuals involved, whether as team members, employees of organizations or teams, or as single individuals needing special attention. Two examples will help clarify this point. When restructuring occurs in an organization, a common happening

today, attention should be given to how the newly connected individuals will interact. Personality profiling and what is known about previous work history of the involved individuals are useful here. Redrawing organizational charts puts different individuals together who may not have worked together previously. This often results in conflicts that slow performance. That can and should be discussed and managed in the beginning of the reorganization.

When providing training in leadership, selling, or customer service, for example, feedback to individuals on their identity and how that will impact their behavior in these roles provides opportunities for improving performance. If a customer service provider discovers that his or her style includes a high degree of assertiveness, he or she can focus on developing listening skills so as to avoid initiating conflicts with demanding customers. More about this in Chapter 15, which concentrates on training.

2. Each individual makes a conscious or subconscious mix of decisions about how hard he or she will work to perform well or to improve his or her performance if that is needed. This seems to have something to do with how individuals see the effectiveness of their own efforts at reaching goals, both their own goals and those of the organization or team. This is called "expectancy," and the well-known concepts used to describe its influence on individual performance is Expectancy Theory. "Expectancy relates to the probability of an outcome occurring." (1)
3. The best way to positively impact motivation is to find the connection between the personal goals of the individual and the goals of the organization or team. All individuals have goals, even if they are poorly defined. Goals come from our social and physical needs, as discussed frequently. Because the nature and strength of needs vary between individuals, goals that people seek to satisfy their needs also vary in their strength or importance. Thus, the best way to affect motivation is to identify the individual's needs and his or her strengths. Then the key is to establish an environment in which achievement of the organization/team's goals gives the individuals what they want. Figure 13.2 shows this clearly.

Figure 13.2 The Model of Motivation.

Encouraging motivation among employees is among the most common concerns identified by leaders of organizations or teams. However, they tend to look for a single magic answer, something that they think they can do that will work for all individuals. The best approach is to identify the *different motivations* for each employee, and then seek to provide goals, rewards, and recognition that meet those motivations. That is how the High Achieving Managers, the small percentage of successful managers discussed in a previous chapter, handle motivation.

Identifying the motivations of employees is not as complicated as it sounds. People will identify their own motivations when asked. In addition, employees often discuss the things that are important to them — what they want in life or at work, sometimes in casual conversation. Managers who want to can find ways to identify the motivations of their people. As one of our most astute leaders likes to say, "Find out what is important to the employees."

The biggest challenge for managers is to find ways to meet the motivations of others once they know what those motivations are. Some people want tangible rewards, so pay and promotions are a key. Others want recognition for their work, so providing sincere, specific, and timely feedback on performance is important to their motivation. Still others want challenging assignments or variety in their work, while their co-workers want stability and predictability in assignments. The challenge is to meet the individuality of the employee.

THE MODEL FOR GROWTH AND DEVELOPING INDIVIDUAL PERFORMANCE

As discussed, the Model for Growth (Figure 13.3) clarifies methods for developing individual performance in a number of ways. If one teaches a team of success-oriented salespersons how to maintain positive attitudes through conscious use of affirmation and visualization (which we all do without thinking about it), it may well enhance identity and then improve sales performance (behavior). If a team of curious managers decides to improve team problem solving by getting everyone involved in a seminar to learn teamwork skills and then follow up by setting weekly team meetings, they may have increased their knowledge and their skills. They will have also changed behavior as they reschedule workloads to make time for the team meetings. Finally, if they make progress from the team meetings in identifying and solving real issues blocking goal achievement and thereby improve performance, the curious managers have found a "new" way to do things.

The Model for Growth is most fundamentally a set of *six techniques for the individual to take responsibility for his own performance improvement*. If one looks at the model from front to back, one can see that developing

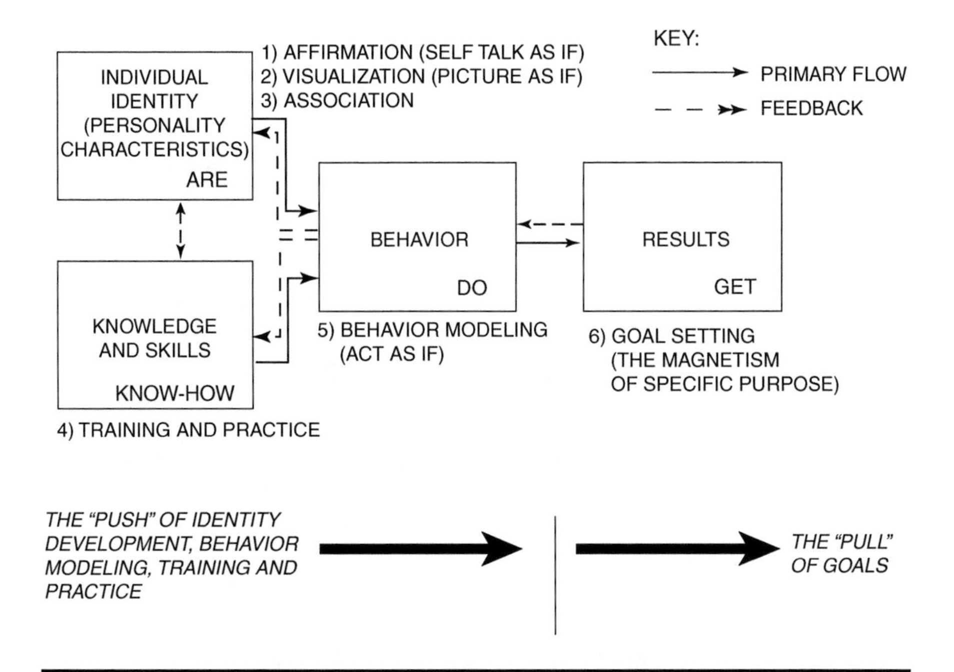

Figure 13.3 The Model for Growth.

one's self has to do with setting clear goals and finding ways to adjust behavior to meet these goals. This is particularly important for the highly motivated and often impatient person wanting immediate success. Goals work directly on performance results; and if the goals are successfully met, the highly ambitious and impatient person gains quick satisfaction. For this type of person, goal setting can become very addictive.

The Model for Growth also reminds us of the power of learning from others, finding desirable models. It suggests that individuals find acceptable models of effective performance to imitate as they stretch to reach their performance goals. A model of effective salesperson and one for the High Achieving Manager are only two examples. Knowledge and skills in the Model for Growth remind us that learning is a lifelong commitment, as changes occur and in fact accelerate in technology, the economy, our organizations or teams, and therefore our jobs. Finally, the Model for Growth is a reminder of the importance of one knowing oneself and how to get the most out of who one is while working on who one wants to become.

To some, the Model for Growth is suspect because it assumes that individuals *can* improve their own performance, and that they *may want to do so*. Developmental psychology provides some insight into whether individuals can develop themselves, sometimes with organizational or team support in an occupational setting. One recent excellent book on developmental psychology provides the following summary of basic issues regarding developing the individual. (2)

As that publication noted, the question of nature versus nurture has long been a focus of debate. For example, are leaders born or are they developed? "On the nature side of the debate have been those who emphasize the influence of heredity, universal maturational processes guided by the genes, and biologically based predispositions." However, "On the nurture side of the debate have been those who emphasize environment — forces outside of the person, including life experiences, changes achieved through learning, and the influence of methods of child rearing, societal changes, and culture on development." (2)

The authors of that textbook argue that most "developmentalists" today see individual development as an "interplay" between both nature and nurture — between who we are at any point in our lives and the environment in which we live. (2)

The basic question is whether we as individuals can take an active role in our own development by planning our learning and life experiences. Or are we destined to be passive recipients of the influence of forces beyond our control? This basic question is a complex one. Like other questions of great complexity, there may be no final answer. Theorists disagree about how active individuals are in creating their own environments and, in the process, producing their own development.

Because the experts do not agree on the "activity/passivity issue," each of us gets to decide whether to sit back and let the forces have their way with us, or whether we will make efforts at influencing our own world so as to develop in ways we prefer. This is not a question reserved for philosophical debate. For example, many managers seeking help in improving their leadership and management style wrestle with the question of whether they will take responsibility for improving their performance in the role of leaders and managers. The alternative is for them to continue in what they are doing, although if they sought help to change, it is usually because they have some evidence of the need to change how they manage to improve performance results. Some managers decide to make the effort to change. Others decide that they cannot learn better ways, or that it is not worth the effort, or that the organizational or team culture is too powerful an influence on them. Whether these leaders and managers decide to work on some changes in their managerial style and accomplish performance improvement, or decide to stay the same because the results of their managerial style will not change.... "either way, they are right."

The book referenced above, and others in the field, argue that development is a lifelong process, not one occurring only in the early years of life as psychology a few years ago maintained. While the subtleties of how we develop and how much we can consciously influence our development are a matter of debate, it is clear that in today's world "one is never too old to be a developing person."

MORALE AND INDIVIDUAL PERFORMANCE

Morale is the set of attitudes employees have about their working situation, and it impacts performance in a number of ways. Developing the individual and his or her performance sometimes involves figuring out what morale is like and identifying and acting on morale where it is deficient. Morale tends to be the result of attitudes about three areas of the work situation:

1. *Employee attitudes about the organization* include a number of areas of common concern to employees at all levels. These concerns usually include at least the following: how the organization treats its people, and whether the organization does good work and provides a desirable quality of products and services. Attitudes about the organization also usually include how well the organization is managed.
2. *Attitudes about one's boss* is a major source of morale, good or bad, for most employees. This can range from adoration of the boss to the more frequent contempt for the boss because of poor treatment. Questions or topics of discussions about bosses tend to be in the following areas: the boss is too demanding and has little tolerance for mistakes, or, alternatively, the boss is too easy — it is like not having a boss at all. Maybe the boss treats people fairly or has a few favorites. Sometimes, morale focuses on the boss providing support and recognition for top job performance or totally ignoring employees and what they do.
3. *Attitudes about one's job* are also a source of morale that is often overlooked by organizations seeking to understand this organizational component. Issues include the following: the job is fun and enjoyable, or going to work is a pain; the employee feels competent at what he does, or realizes there is a lot he is not good at. Maybe the morale issues focus on the job being too routine or providing very little empowerment and decision freedom. Positive morale in this regard will have to do with the job providing the amount of empowerment preferred by the employee, which varies to some extent with the individual and his confidence.

Actions for improving morale begin with identifying the areas in which morale is an issue. As examples:

- Establishing and communicating a clear strategy can significantly improve morale within the organization and how well it is managed.

A lack of strategy or understanding of the direction of the organization is frequently a source of confusion and poor morale.

- Establishing a performance management program ("appraisal system") with an employee development component can help morale in all three levels of morale identified above — those about the entire organization, the boss, or the individual's job.
- Working to improve leadership style and management skills in the organization or team goes a long way toward improving morale about one's boss, as long as the boss is trying to improve in a sincere way. Conscientious efforts at improvement by managers gain a lot of credit with employees who see their bosses working to improve style. Actually improving leadership and management style is also received well by employees. One employee, himself a manager, made the following comment recently in a management development program: "I know why we are doing this. My CEO is in search of a leadership style. I respect the effort."
- Setting team goals and building team spirit can help in all three levels of morale and provide the opportunity for individuals to ask for help in their own performance from other team members.
- If negative morale is widespread, organizations need to look to their hiring process as one possible cause. First, any organization demands certain types of performance: fast paced versus slower paced, dedicated customer service or not so important, etc. People have to want to work in the culture of the organization. The organization can do a lot in the hiring and orientation of new employees to improve the fit of the individual to the specific job competencies and organizational culture. This helps ensure that employees can become competent in their jobs and feel good about their work. In addition, work designed to improve job enrichment may be needed for individuals bored with their current work.

PERFORMANCE MANAGEMENT AND DEVELOPING THE INDIVIDUAL

As discussed in Chapter 11, a very effective way to work on developing individual performance is through a well-designed performance management program. This is one of the most common organizational components, with close to 90 percent of organizations using this method to review performance. Still, they are frequently done poorly. Performance management has great potential for developing individual as well as team performance because it can serve as a primary link between organizational or team goals and those of individual employees. Chapter 11 discussed this performance improvement method in detail, including specialized techniques such as realistic job previews. Here we provide an overview of what is needed to make this initiative effective in developing individuals.

Remember that some management gurus and more than a few organizational leaders have negative attitudes about performance management programs. (3) Experience shows that performance management programs can be very effective in developing individuals if they follow these guidelines:

1. Clear statement of organizational or team goals is a prerequisite for performance management. That way, it is possible to set individual performance direction and have an objective basis for evaluating individual performance.
2. There should be clear identification of the core competencies needed for the job being managed. This is one of the most difficult guidelines to meet, but is achievable if the organizational leadership is dedicated.
3. The performance management system should include a collaborative method for development of individual performance goals and key performance indicators (KPIs). This means that both the manager and the employee collaborate on identifying performance criteria ("behavior" in the Model for Growth) and goals ("results" in the Model for Growth).
4. Another very difficult guideline for organizations to achieve is to have and use methods for regular feedback to employees on their performance. Accomplishing this guideline is difficult for a number of reasons, including managers' anxiety about their ability to coach and counsel and the increasingly rapid pace of work in today's organizational world. Time is at a premium in almost all organizations. Still, how much time is spent fixing problems that some earlier individual coaching might have avoided?
5. The organization should provide training and development opportunities for enhancing the knowledge and skills of the employee that are supportive of his or her achieving a high level of performance.

A review of a brief case study discussed in a previous chapter will illustrate good use of many of the guidelines stated above.

Organization: Pro*Visions Pet Specialty Enterprises Sales Team for Ralston Purina Co.

- Products and services:
  - Dog food
  - Cat food
  - Specialized pet foods, litter

- Markets:
  - Pet superstores
  - Regional chain stores
- Primary issues:
  - The need to clarify its goals and actions in support of Ralston's corporate strategy
  - The need to build the team toward increasing cohesion and performance

The team had received corporate objectives and goals for the coming year. Having a good deal of independence in how to achieve the team's part of the corporate strategy, the team leader decided that the group should develop a strategic plan of its own, remembering that its own strategy should support that of the overall corporation. The team concluded strategic planning by developing a group of strategic objectives aimed at developing team performance. Following that, the group decided to drive the strategic objectives through each of the sub-teams within its larger team. The group wanted each sub-team to have accountability for identifiable sections of the strategic objectives and for individuals also to include their specific goals in their individual PMPs. The group felt the need to conduct a process to cascade these objectives down to the individual level, rather than leaving it strictly up to the sub-teams and their leaders and employees. Busy schedules dictated that the Pro*Visions team be involved in making sure the process of goal setting was completed by all individuals in their group so that performance would be efficient and coordinated.

After each member of the Pro*Visions leadership team developed plans for his or her sub-teams, the total group met once more to coordinate all the efforts involved in their group plan. The steps they used were as follows:

- Each team member listed his or her individual action plans, which included both strategic objectives and the goals for daily sub-team operations.
- To the extent appropriate, the team members also discussed the goals and activities of the employees they managed where these goals and activities had significant impact on those of the Pro*Visions group.
- The total team reviewed and discussed individual action plans looking to coordinate efforts, identify gaps in what needed to be done, and eliminate duplication. Further conversations with the Pro*Visions leader were scheduled for detailing of plans and identification of support needed by the managers and their sub-groups.

- The Pro*Visions leader also identified training and development services needed for his team members and those they managed. To a significant extent, those training opportunities were tied to the immediate performance objectives developed by this total team planning process. This included the installation of hiring tools and feedback on management style.

CORE COMPETENCIES AND DEVELOPING INDIVIDUAL PERFORMANCE

Developing individual performance is helped significantly by recognizing that people are influenced in their job roles by what is occurring in their personal roles. While some employees are "private" and keep things to themselves, others appreciate and benefit from training in areas such as learning better communication, goal setting, and time management skills, which are skills that can be applied both at work and in personal situations. Others appreciate learning methods for improving personal finances and make use of employee assistance plans.

But realistically, organizations provide these "personal" support programs primarily because they hope that a successful or "less-troubled" personal life will lead to better job performance. To help ensure "better job performance," organizations need to know what it takes to perform well on the job and then communicate that to the employee. The question organizations should ask and answer is: "What are the core competencies for successful performance for the various positions?" Very briefly stated, knowing core competencies has the following benefits:

1. They provide the basis for rationalizing the hiring process so that people who have the core competencies for each position can be identified.
2. Identification of core competencies can drive the training and development plans so that instruction provides help to those seeking to improve performance through developing these competencies.
3. Core competencies provide a basis for setting performance goals at the beginning of the performance management cycle and then evaluating performance at the end of it. They also identify target subjects for coaching and counseling individuals during the performance management cycle to develop people's abilities in the core competencies.

Dubois defines job competency as the "underlying characteristic of an employee (i.e., motive, trait, skill, aspects of one's self-image, social role, or a body of knowledge) which results in effective and/or superior performance in a job." (4) The wisdom in this definition is that while it allows for the fact that the requirement for specific competencies comes from the nature of the work (job tasks and activities, roles, and organizational environment and issues), the competencies lie within the individual. The key point here is that once job competencies are identified, they can become the basis for hiring and then developing individuals toward those competencies.

CONCLUSION: INDIVIDUAL LEARNING AND PERFORMANCE IMPROVEMENT

As noted at the beginning of this chapter, all performance improvement comes back to developing individuals, at least in part. Individuals ultimately must learn how to make the strategy happen, or the new structural design work. Or they must learn how to sell the new product and manage the changed production process. Those of us working in performance development often get enthralled with new performance improvement interventions and forget a basic truth: when the interventions work, they work because human beings have the attitudes, the motivation, the knowledge and skills, and whatever other competencies are needed to make them work.

SUGGESTED ACTION STEPS

1. Identify the common personal issues that impact performance in your organization or team (financial problems, chemical abuse, etc.). Take actions to establish programs as needed.
2. Review any available data on morale in your organization. If data is sorely lacking, conduct a brief morale survey. Decide what areas of morale need work, and what sources of poor morale exist. Take action to eliminate or reduce those sources.
3. Review your performance management system. Is there sufficient goal setting and sufficient feedback on a regular basis regarding performance? If not, take action to build those performance management components.
4. Make sure that your understanding of the job competencies for core jobs is complete and up-to-date. If not, action is required.
5. Make sure that the core competencies are used in your hiring process.

END NOTES

1. Smither, Robert D., *The Psychology of Work and Human Performance,* Longman, 1998, p. 217.
2. Sigelman, Carol K. and Shaffer, David R., *Life-Span Human Development,* Brooks/Cole Publishing, 1995, pp. 29–30.
3. Cummings, Thomas G. and Worley, Christopher, G., *Organizational Development and Change, sixth edition,* South-Western College Publishing, 1997.
4. Dubois, David D., *Competency-Based Performance Improvement: A Strategy for Organizational Change*, HRD Press, 1993, p. 9.

Chapter 14

LEARNING IN PERFORMANCE AND PERFORMANCE IMPROVEMENT

THE IMPORTANCE OF LEARNING IN ORGANIZATIONAL PERFORMANCE IMPROVEMENT

Many books, articles, and seminars on organizational development, training and development, and developing leaders and managers focus on programs as if they were discussing the construction of a building. They focus on the "foundations" of the program, structure of the suggested actions, and related skills. For example, the index of one of the most complete and useful books on organizational development has one reference under "Learning." (1) To be accurate, this well-written compendium discusses learning as an element of its numerous topics. However, as is usually the case with subjects in these areas, learning is not presented as a critical success factor for organizational performance and performance improvement.

Our decades of experience in facilitating performance improvement programs has led us to believe that participant learning is the most important element in performance and especially in performance improvement. We use the term "learning" here in a very broad way, meaning a significant change in someone's thinking or behavior. To help organize the focus of the following discussion, the reader is reminded of the Model for Growth (Figure 14.1).

If one focuses on the *identity* of an individual, then one recognizes that there can be many areas of change and learning in that individual's identity. Because life consists of a series of developmental stages from womb to tomb, each developmental stage requires learning.

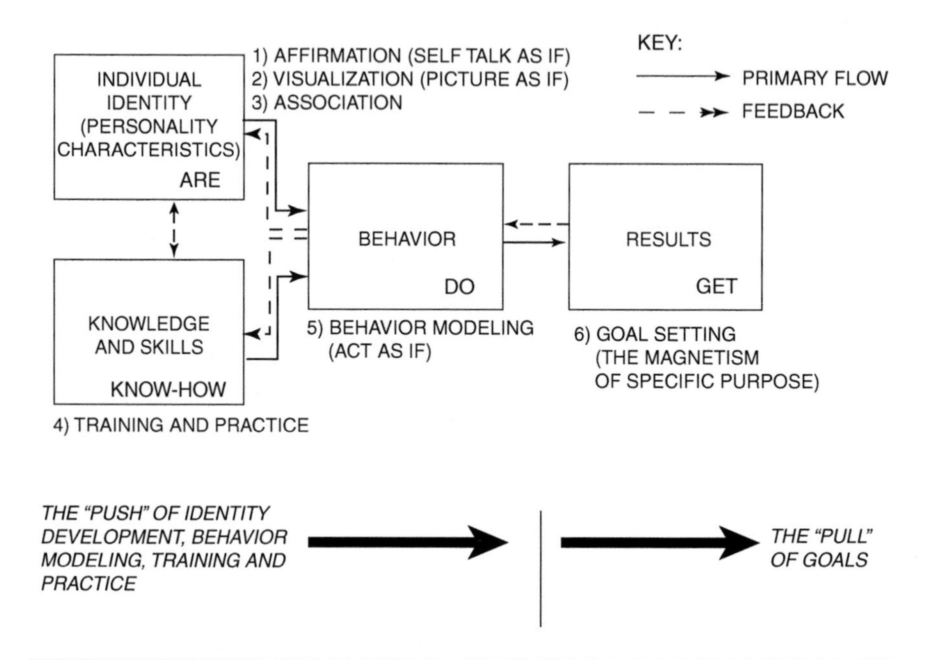

Figure 14.1 Model for Growth.

> "To grasp the meaning of life-span development more fully, we should also understand two important processes, maturation and learning." ... "The second critical developmental process is learning, or the process through which experience brings about relatively permanent changes in thoughts, feelings, or behaviors." (2)

For those of us working in organizations, that is part of our life experiences. When our organization undertakes performance improvement efforts, that is part of our experience as well, particularly if we are an active participant in the improvement process. Any number of the aspects of our identity can change as we have these experiences. We might well learn some concepts, methodologies, or suggested behaviors and skills from the organization's performance improvement programs.

Two other aspects of our identity might also change as we participate in a performance improvement program. Our motivation can be impacted, either for the better or for worse. We might, for example, decide that the performance program is poorly targeted and reflects our leaders' lack of awareness about the "real problems" in our work situation. Or, we might decide the program is directed at areas that need improvement, so we become motivated to participate. If we decide we like the direction of the performance program, our attitude about what is being done, and maybe our attitudes about the boss or the company, can also be affected. If we do not like a program aimed at performance improvement, then it becomes "one more dumb idea around here." When employees or managers are involved in an organizational program, they decide how much they will participate. Making that participation worth the effort is part of what the facilitator should have as a goal for the program.

The primary value of what is learned during a performance improvement program — concepts, skills, techniques, processes, etc. — comes from its use in the organization. How much of what a program teaches is used is an interplay of the following factors: the organizational leaders' emphasis on the importance of using what has been learned; whether or not the participants thought the concepts, recommended behavior, and skills were worthwhile; whether the participants learned enough concepts and skills to know what to do; and whether they will see some fairly quick results from their efforts at getting the new skills into their behavior. In short, will those working with new knowledge and skills see quick results? All of these factors are important in whether the learning is applied and whether it becomes permanent. Our experience says that leadership support is the most important among these factors.

The concepts, techniques, systems, procedures, and skills that are part of a performance improvement intervention or program should be provided in a way that encourages learning. This is clearly the case for a program that is primarily a training and development intervention, such as teaching people supervisory and management concepts and skills, or teaching people how to sell or provide customer service. However, the same learning must occur in other performance improvement programs that are not primarily training and development efforts. For an organization and the participants involved in any performance development program to make good use of it, they must learn the important ideas and skills to make the new strategy, hiring process, organizational structure, or Six Sigma system work. In many cases, the original participants in a performance improvement intervention are viewed by the organization as only the beginning, the "pilot group." In this case, spreading the performance improvement effort further becomes the obligation of the organization.

What follows is a discussion of some of the primary ideas, skills, and attitudes that must be learned in each of the major topics discussed thus far in this book. Also discussed is learning about one's own motivation to learn where that is appropriate.

LEARNING AND PREVIOUS TOPICS

1. Chapter 2 discussed many topics about getting to know an organization and its performance status and improvement needs. A central concept was to focus on where the organization needs to be, defined primarily by organizational leadership. Then a review of the current status in various areas (e.g., employee turnover rate or service ratings from key customers) must be completed.

   The ability and motivation to assess one's organizational performance is an approach that often must be learned. This is particularly true of the ability to "confront the brutal facts," and to use data where possible to clarify, confirm, and amend one's own perceptions. (3)
2. The use of a systematic approach to performance improvement, such as the General Model of Planned Change (GMPC) discussed in Chapter 3, is lost on many organizational leaders. Often, leaders are too impatient to adhere to an organized series of steps in assessing their organization and learning how to manage and be involved in performance improvement efforts and programs.

   One reality that can be used here to help leaders and managers develop persistence at learning and following the GMPC is the importance of their support for achieving performance improvement. One way to get anybody motivated to learn and act in a needed way is to make a convincing case about his or her importance for success in the endeavor. It is remarkable how fast people can learn to do something they have decided is really important and something about which they really care.
3. Perhaps the greatest learning challenge is for people starting new organizations. This demand for quick learning of ideas, techniques, skills, and new behaviors is high whether the organization is a brand-new one, a new department in an existing organization, or a dramatically restructured organization. Often, the managers, leaders, and employees need to learn what to do for the first time; or they need to learn how to forget and stop doing what they have done for many years, as in the case of a restructured organization. The challenge is to learn rapidly in any new organization, whether brand new or re-made.

   There are a number of aids in learning how to create or run a "new" business. Professionals such as corporate attorneys can help with the legal structure and the pros and cons of various legal formats. There are also an increasing number of colleges and universities teaching what is referred to as "entrepreneurship" to help those involved in leading new organizations.

4. The learning involved in goal setting, the primary topic discussed in Chapter 5, is primarily skills in goal development and time management. But to do this, individuals must come to some understanding of their own needs and how goal setting, personally as well as professionally, can add to their success in life.

   Most people do some kind of goal setting at work, but less than half of all people in the United States have written personal goals. Part of what people can develop is an appreciation for the personal payoffs of having working goals. In addition, individuals can develop skills in setting their goals professionally, personally, and financially. They can then develop daily habits of attaching time management to their actions related to the appropriate goals that day, week, or month. The motivation for those not practicing goal setting to begin to do so is sometimes the result of the recognition that written goals are dramatically more apt to be completed than those in our heads but not written down. Perhaps more significant is that when a specific goal is part of a large general plan, the goal carries increased importance to the involved individual. So, creating an overall goal plan for life shows the individual more about the importance of individual goals.
5. Understanding strategic planning must occur at two basic levels for the individual to be successfully involved in the creation of the strategic plan or its implementation. First, the person involved in creating the strategic plan must master the basic concepts describing the process. Driving force, as discussed in Chapter 6, is not a simple concept to understand. Deciding which driving force best serves a person's organization is even more difficult for most people. In short, the concepts that define each of the four or five stages in strategic planning are subtle.

   Second, the process of designing a strategic plan for the organization where a person works can be intimidating and sometimes creates hardened attitudes that restrict creativity. Creativity can be increased to some extent by developing a confidence about one's own ability to be creative. In addition, instructing someone to be creative and to avoid obvious answers to a problem increases creativity. In general, "most of us have creative potential, if we can learn to unleash it." (4)
6. It may seem strange to discuss the role of *learning how to create and use a learning organization* along the lines of the discussion in Chapter 7. However, the efforts at creating learning organizations require more than having training programs. As is apparent from the organizations discussed as learning oriented in Chapter 7, building a learning-oriented culture requires a multifaceted

approach, from hiring to finding ways to store and provide access to valued data, concepts, and other types of knowledge. It also should include assessing the job applicants' motivation to learn. Finally, it is known that recognizing and rewarding a behavior significantly increases the occurrence of that behavior. Therefore, recognition and rewards should be tailored for people who accomplish learning and make use of that new knowledge. All of this requires some major changes in the standard way organizations treat knowledge and learning. Thus, people throughout the organization must learn how to create and use the new learning systems and the knowledge that they contain.

7. Learning to hire effectively is straightforward compared to figuring out how to make the organizational culture more learning oriented. However, many people, especially managers, continue to do this poorly. The key to getting people motivated to learn how to improve their hiring steps and then to use them requires two things: leadership must insist on improvement in hiring results, and people should be made aware of the costs of poor hiring. Once people become motivated to improve their hiring involvement, then learning how to use a rational hiring process and the applicant assessment tools discussed in Chapter 8 is comparatively easy.
8. Learning effective leadership is probably more complicated than learning to hire effectively. As is always the case, learning the concepts and understanding the associated techniques for leadership are easier than learning to systematically use the concepts and techniques. Thinking back to the Model for Growth, "it is a far distance from knowing (identity) to doing in your behavior." When people seek to make use of the Achieving Manager as described in Chapter 9, understanding and becoming competent in all six areas of management is a challenge.
9. Learning the concepts describing what elements comprise an effective work team is straightforward and can be quickly understood and applied. Figuring out when to use teams and when teamwork is not indicated takes more work and experience in using a model such as what was discussed in Chapter 10. An individual's personality, part of their identity, plays a part in his or her willingness to make use of teamwork. Some people prefer to work alone; others are more naturally team members.
10. Performance management is also an area where some managers struggle, largely because of minimal motivation to do performance management programs (PMPs) at all. But even when managers or employees are willing to engage in PMPs, they must learn to use good listening skills. The most difficult thing to learn in performance

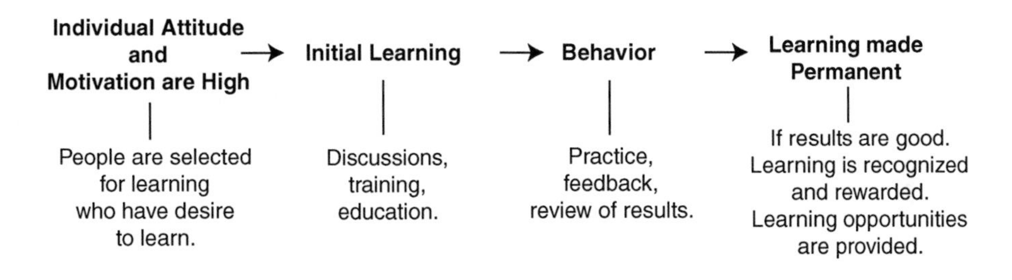

Figure 14.2 How people learn.

discussions is how to treat feedback as a professional, not a personal, comment. Some managers sound critical or appear to be attacking when providing feedback to their employees. Employees must learn not to take feedback defensively and to use that feedback constructively.

More specific areas that must be learned for an effective performance management process start with the techniques for goal setting. A related area of learning is the specifics of any behavior style factors that are included in the reviews. For example, many companies or universities use "responsive customer service" as a behavior measure for customer service providers. Tying that phrase to specific behaviors takes some effort and learning to be valid as feedback.

A PRACTICAL MODEL OF HOW PEOPLE LEARN

Part III discusses systems for helping people learn as an organization-wide program. Developing employees at all levels of the organization is a leadership obligation. Leadership support is an absolute requirement for the establishment of a learning culture. What follows is a brief discussion of how people learn new ideas, attitudes, behaviors, and skills.

SUGGESTED ACTION STEPS FOR ORGANIZATIONAL OR TEAM LEADERS

1. Identify the areas of your business where knowledge, skills, behavior, and appropriate attitudes and motivation are essential to organizational success.
2. Establish a process for identifying the specific areas of knowledge, skills, etc. indicated by your answers to action step 1 above.
3. Identify methods for closing the gaps in knowledge, skills, behavior, motivation, and attitudes indicated by your work in action steps 1 and 2.

END NOTES

1. Cummings, Thomas G. and Worley, Christopher G., *Organizational Development and Change, eighth edition,* 2005, see p. 686, Subject Index.
2. Siegelman, Carol K. and Shaffer, David R., *Life-Span Human Development,* Brooks/Cole Publishing, 1995, pp. 3 and 4-ff.
3. Collins, Jim, *Good to Great,* Harper Business, 2004, see Chap. 4.
4. Robbins, Stephen P., *Organizational Behavior,* Prentice Hall, 1998, p. 105.

Chapter 15

PROGRAMS FOR TRAINING, DEVELOPMENT, AND EDUCATION: HELPING PEOPLE LEARN FOR PERFORMANCE

INTRODUCTION

People working in organizational development or performance improvement often use the terms "training," "development," and "education" more or less interchangeably. While these are distinguishable performance development initiatives, they share a common purpose, which is to provide participants with concepts, techniques, motivation, and skills that can enhance performance now or in the future. In various ways, these three learning initiatives seek to inform us, provide us with concepts with which to work, teach us skills necessary for performance, and encourage positive attitudes and motivation so that we will use the concepts and skills successfully.

DISTINGUISHING TRAINING, DEVELOPMENT, AND EDUCATION

These three terms are used in various ways in different organizations. Because so much effort and resources are put into the three areas, clarifying what each means can help organizational leaders in deciding how and where to use these learning initiatives. The organizational use of these interventions is enormous. In early 1998, the ASTD estimated that all U.S. organizations spent a total of $55.3 billion on training in 1995. (1) This expenditure probably did not include the salaries of participants in training or the cost of some educational programs such as college tuition or executive development seminars.

1. *Education,* in terms of current usage in organizations, is the broadest of the three terms. Most commonly, the term "education" is used in two ways by organizations. First, education is used to mean everything an organization does to support learning, including job training, executive development, and tuition support for college courses. Second, the term is used to mean an organization's support for its employees attending college. In the second more restricted use of the term "education," training and development is then seen as everything else the organization does. (2)

   Many organizational and team leaders have found the following use of the term "education" helpful in planning their learning programs. Education includes:
   - Methods and programs for learning knowledge and concepts believed useful for the long-term performance or career advancement of individuals in the organization
   - Learning about performance programs or initiatives believed beneficial to the long-term performance of the organization or team. Examples include:
     - A seminar for organization or team leadership on the characteristics and strategy for establishing a corporate university
     - A workshop for leadership on the approaches, costs, and benefits of process re-engineering and project management
2. *Training* and *development* can be more specifically defined than the common use of "education." The Identity/Role theory, discussed in Chapter 14, helps here.
   - *Training* is teaching people knowledge, skills, and attitudes about the roles they play, or will soon play, in the organization or team. Examples are as follows. A salesperson is taught techniques for closing business, prospecting, or converting a customer's resistance to the purchase of the product. Managers are taught the importance of and techniques for goal setting. Customer service providers are taught how to recognize and deal with various types of hostility in customers. These examples of skills training all have to do with a specific role the individual plays in the organization. Some skills, for example, goal setting, can be used in more than one role of the learner.

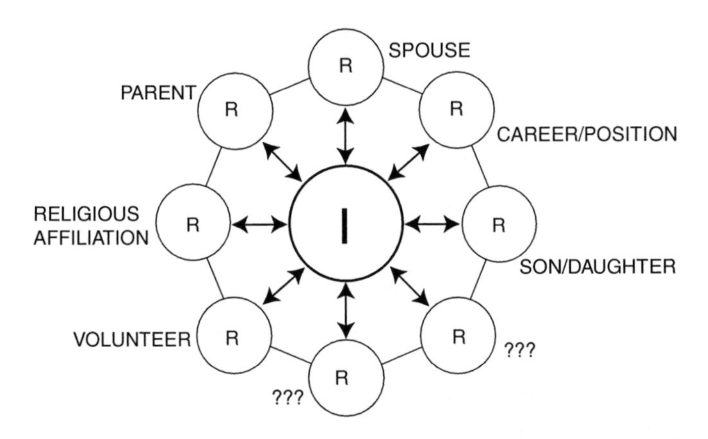

Figure 15.1 Identity/Role Theory.

- *Development,* on the other hand, is helping individuals discover or better understand their own identities and how to get more out of who they are. As discussed in Chapter 13, it also includes belief in and efforts at developing one's personality, knowledge, and skills in multiple roles, motivation, and capacities. Review of a second simple model will further clarify the distinction *and* relationships between training and development.

  Performance = Clear Goals + Good Attitudes + Knowledge/Skills + Motivation

  Very simply, development has to do with work on the identity factors of goals, attitudes, and motivation. Training focuses, for the most part, on improving knowledge and skills related to one's job roles.

THE THREE LEARNING INITIATIVES AND PERFORMANCE IMPROVEMENT

A number of excellent books have been written on the critical connection between corporate strategy and training, development, and education for an organization. (3) The devil, however, as they say, is in the details. Specifically, there are at least two major questions here:

1. Which of the knowledge and skills areas are most urgent for reaching our strategy? What do we need to learn to make the strategy happen?

2. What learning initiative best fits with other performance improvement initiatives we are using, such as process re-engineering, organizational restructuring, etc.?

Chapter 14 focused on learning specific concepts and skills pertaining to ten performance improvement initiatives, such as strategic planning, hiring, and project management. This section of the current chapter focuses on the overall relationship between learning and the organizational strategy and operations. The conclusion of this chapter focuses on what we know about techniques for effective training and development.

Strategic planning and training and development are connected in two separate but related ways. First, the best benefit to the organization or team involved in strategic planning is not only to get a written strategic plan, but also to learn how to think strategically and do planning. The first time the top team strategizes, it may need a facilitator from outside the group to learn how to do this type of planning. If training the team in how to do strategic planning is intertwined with the process of actually developing the strategy, the team becomes more independent and more knowledgeable of the process it has experienced. The team learns to plan.

Developing individual performance, discussed in Chapter 13, has many connections to training and development. In today's organizational world, there are fewer middle managers and increased demand for speed and effectiveness in dealing with customers and clients. As a result, individual employees need to know how to self-develop and self-manage. This is one of the key points in Drucker's recent book. (4)

There are issues occurring in the United States that make the idea of a highly developed employee group capable of self-management a major challenge. There is evidence that the middle class is shrinking as a percentage of the U.S. population. In essence, we may be becoming more of a two-class society, those at the top of the economic ladder and those toward the bottom of it. While generalization is dangerous, it is essentially true that lower income neighborhoods and families have inferior educational systems compared to the middle and upper income groups. School dropout rates are also higher among lower income families. The debate about what causes this and how to fix it is beyond the focus here. The point is that corporations often have to do more training of employees than in previous years. This is partly due to growth in the complexity of work because of high technology demands and the increase in competition between organizations requiring them to develop customer service and other skills in their employees. The former U.S. Secretary of Labor, Robert Reich, summarized this issue succinctly: "At least one third of the current workforce continues to be unprepared." (5)

As noted above and in Chapter 14, learning is a constant requirement for today's labor force. That means that training, education, and development are part of both performance development interventions such as strategic planning and also independent performance improvement requirements. These learning opportunities are used and needed when individuals require additional knowledge and skills for job performance, whether or not another improvement intervention is underway.

CHARACTERISTICS OF EFFECTIVE TRAINING AND DEVELOPMENT

Because training and development are valuable performance improvement interventions — both as a stand-alone activity and also to support other interventions — what does it take to make sure these efforts work? Why do so many training and development programs seem ineffective, and how do we overcome that ineffectiveness? (6)

1. Those cited previously in this chapter who stress the importance *of tying education, training, and development to the strategy of the organization/team* are totally correct about the importance of doing that. Again, the primary issue is to identify exactly what learning outcomes are needed most urgently and which ones best benefit the most important elements of the strategy and the person(s) who will undergo the learning. One cannot do everything at once to improve performance, but in today's competitive world, one must always be working on something. The best approach for managing learning required strategically is to make it part of the strategic plan.
2. The role of *dedicated and persistent leadership support is crucial* to making any learning initiative more than an academic experience. People in a position of organizational leadership enhance learning delivery programs by stressing their importance to current or future performance. This is critical to the success of learning programs, both while the programs are occurring and in ensuring the learning is applied back on the job.

These first two characteristics of effective training — tying it to strategy and getting leadership support — are closely connected. Creating and pursuing the organization's strategy is a primary responsibility for leadership. Dedicated leadership can start their support of learning by making sure the strategy is clearly defined and that employees throughout the organization know the essence of the strategy. This summary of the organization's strategy throughout the organization can include a strong statement regarding the required knowledge, skills, and attitudes supportive of, and usually included in, the strategic plan.

As a well-written book puts it: (7)

Learning Experience × Work Environment = Business Results

It is natural to see learning as more worthy if it is connected to organizational strategy. It is even more natural to see learning as worthy if the leadership of the organization shows how it serves the best interests of the organization and those who help the organization be successful.

3. *A well-defined and facilitated learning experience* (education, training, and development) is an obvious, but sometimes elusive, requirement for helping make learning experiences beneficial for performance. This characteristic starts with the learning facilitators' dedication and persistence to doing what it takes to provide the requisite learning opportunities in the training program. The requirement for making learning effective in improving performance is not the sole requirement of the facilitator, nor is on-the-job application solely the responsibility of the organizational or team leaders. It requires collaboration between the leader and the facilitator. The facilitator should be dedicated to helping the learning participants see how the knowledge and skills can be implemented at work. The learning program should also contain numerous activities that require efforts in that implementation. Using real work examples, either provided by the learning facilitator or provided by learning participants, is one technique for making the learning "real life."
4. *Awareness of the role of individual identity, and how it impacts an individual's performance in even the most technical roles,* is a critical characteristic of effective learning. Participants get weary of learning what appears to benefit only their job role performance. As stated above, seeing ways to implement learning on the job is important for learning to be retained. Significantly, participants are even more motivated by learning that involves benefits for them in areas outside their current work. For example, teaching salespeople how to confront resistance and sell the organization's products or services more effectively may be enthusiastically received by a novice salesperson. However, that learning is usually received even more enthusiastically when some of the skills pertain to learning to confront issues with friends or family. One participant in the State Department of Family Services project described in Characteristic 8 below was particularly impressed with the method of "asking and listening" when she tried it on her teenage son and found that it worked. She decided the techniques were worth using at work as well.

Other elements of the learner's identity are important factors to consider in the design and delivery of learning programs. Details of this requirement for effective training, development, and education are explained further in effective learning technique 9 discussed below. Here we will make just a few points about the importance of individual identity in learning. Some people are visual learners, and need to "see" what they are learning. Some people learn inductively, building from facts or details toward general conclusions. Others want to see the "higher-level conceptual framework," and then move to specifics in a more deductive approach.

An additional brief case study from an organization discussed previously helps illustrate this point regarding how attention to personality, motivation, and attitudes can impact the success of training, development, and learning.

Organization: Veterinary Sales Team, Ralston Purina (now a part of Nestle)

- Products and services:
  - Pet products
- Markets:
  - Veterinarians across the United States
- Structure:
  - National sales director
  - A small number of divisional sales managers with numerous salespeople under their direction
  - Approximately 50 field salespeople
- Issue:
  - The National Sales Director and at least some of the DSMs were experiencing a great deal of interpersonal conflict. Causes of the conflict included disagreement over leadership effectiveness between the National Sales Director, the Divisional Sales Managers, and the field salespeople. Our review of the survey data indicated that people in field sales often felt in the dark about policies and procedures. Communication and trust were at fairly low levels throughout the team. It was impacting performance, and any cooperation between salespeople was severely strained.

It was clear from the interviews we did, along with the survey data, that personal dislikes also played a part in the conflict.

The performance improvement initiative that we all agreed was needed was a leadership development program using the Achieving Manager Model of leadership discussed in the earlier chapter on leadership and teams. The Achieving Manager leadership model includes a requirement for open communication. *This meant that communication would stress managing personal attitudes and biases in addition to learning other leadership and management skills.* Once the need to focus on personal motivations and attitudes became known, a series of discussions occurred between various persons involved in the more personal conflicts. The open communication in the Achieving Manager leadership program provided a valued model for people to imitate. Participants were increasingly motivated to find at least some accommodation between the combatants. This led to an improvement in interpersonal relations between three of the upper-level team members. Effort was also directed at increasing the policy communication with field sales teams, and they were included in future team problem-solving efforts. Some months after the leadership development program, we collected additional survey data measuring the same cultural factors as were measured at the beginning of the project. The data indicated a significant improvement in the attitudes of field salespeople regarding communication between groups of sales personnel. The respondents also indicated a major improvement in their involvement in solving specific issues affecting their job.

This case included specific identity issues such as attitudes toward co-workers, attitudes reflecting personal likes and dislikes, and abilities in confronting and resolving interpersonal conflicts. These interpersonal skills and attitudes are never easy to improve. Interpersonal relations are, however, always present at work, with positive and negative results. Not confronting these personal issues where they exist is to avoid real problems affecting work. It means that learning programs are seen as plastic and ineffective.

5. The use of *spaced repetition* is a key to learning sufficiently to make use of the knowledge and skills. Figure 15.2 illustrates the point.

The vertical dimension going from bottom to top in Figure 15.2 demonstrates retention of learned content. Remember that understanding ideas and concepts is a basis for changing behavior. The horizontal dimension, from the lower left-hand corner to the right side, represents the amount of time that learning and re-learning has occurred. The point of Figure 15.2 is that content heard or seen only once is quickly forgotten over time. Repetition in hearing or seeing that content increases retention over time. When something is retained in the minds and memories of learners, it can have an effect on their performance if they choose to use it. A phrase uttered by learners in our various educational workshops is something close to this: "Now that I see it, I can do it."

Repetition is important for really understanding concepts and ideas and the techniques or skills associated with those concepts. Spaced

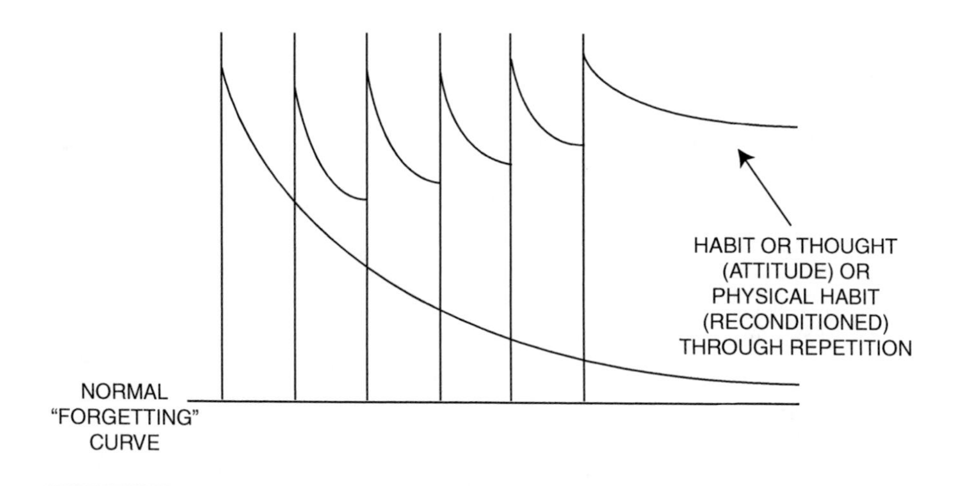

Figure 15.2 Spaced Repetition Model.

repetition is also important for learning to make use of the skills in one's behavior. This is true whether one is learning to use a new technology such as spreadsheets on one's computer or how to differently grip and swing a golf club. Learning to make use of new behaviors, skills, and techniques is accomplished most thoroughly when one learns the ideas and then uses repetition to get them into one's performance and behavior.

A couple of practical considerations in daily training activities are important here. Many participants who regard themselves as quick to learn are especially resistant to any activity involving "repeating of old material." This is sometimes a major issue between facilitators and participants. Repetition and the resulting retention determine how much the participants will be able to use the learning. Happily, in today's high-technology world, it is possible to have learners see the same content through various media, thereby increasing their interest and decreasing their boredom. Reading about, listening to, and seeing the same content graphically is repetition with variety.

6. Effectiveness in training and development is enhanced through the use of a simple method that can be used "on the job." This is very similar to the basic learning model discussed in a previous chapter. Here we are discussing a specific setting: learning in the classroom and practicing back on the job:

Show/Learn → They Practice → Discuss → They Repeat Practice

Here, the distinction between *education* and *training* and *development* discussed earlier in this chapter is important. Education usually involves maximum attention to concepts and minimal attention to skills development. Training and development are much more skills focused; training pays attention to job skills and development focuses more on "personal" skills that can be used anywhere in a person's life. The key point is that knowledge of concepts or skills can only become part of one's behavioral style through practicing those concepts and skills in the real world. While role-playing in a training session can help in the development of "know-how," practicing in the actual work situation is best.

7. Effective education, training, and development maximize *self-discovery*. What that means is that participants should have multiple opportunities to:
   - Decide how the concepts being learned apply to them.
   - Decide how to make use of the knowledge and skills they are learning. It is important to provide plenty of time for participants to set specific goals and plan actions after the learning opportunity has been completed.
8. Education, training, and development should make use of as many of Kirkpatrick's four levels of *evaluation* as possible. (8) Most learning facilitators are satisfied with "reactions" to the learning on the part of participants. As Kirkpatrick argues, participants' reacting well to learning programs is necessary, but not sufficient, for the learning to lead to performance improvement. The following is a fascinating case study of training large numbers of people in a program offering many options for evaluation.

Organization: Division of Family Services (DFS), Boot Heel of Missouri

- Products and services:
  - Welfare cash benefits, childcare, employment training and referrals, healthcare benefits
- Markets:
  - People on the welfare rolls who were under legal pressure from Welfare Reform
- Structure:
  - Area III of the Missouri Division of Family Services
  - Headed by an area director and other central area staff
  - More than 20 southern Missouri counties comprise Area III
  - Each county delivers products and services to those eligible for welfare in their county areas
- Issues and problems:
  - Welfare Reform of 1996 set legal time limits on case benefits to welfare recipients

The Welfare Reform actions taken by the federal government in 1996 have been thoroughly discussed in many publications. It has been shown that welfare reform had huge implications for welfare recipients as well as the national policy and the economy of the United States. (9) Each state has been obligated to meet the time limit responsibilities of the Welfare Reform legislation, although leeway in how they did that existed. Missouri took the initiative, prodded by the Dean from a state college of Health and Human Services, to support a massive training and development program. This program was provided to nine of the county welfare agencies responsible for making "Welfare to Work" happen. More than 250 participants were included in 48 hours of training and development spread over five months of effort.

The evaluation of the program included the following techniques. First, *reaction feedback* from the participants, Kirkpatrick's first level of evaluation, was obtained at the end of each module. We used a learning program critique that was customized to measure the learning content for the work these people did. Second, we tested for knowledge of the training content of a random sample of 10 groups of participants, approximately 25 participants in each group. We conducted measurements of content knowledge in both pre- and post-seminar settings. Some groups were given both post- and pre-seminar content tests. Some groups were measured only pre- but not post-seminar, and vice versa. As a control measure, the content learning of a few of the groups was not measured at all. Basically, the goal was to identify what the participants learned from the sessions. This was use of Kirkpatrick's second level of evaluation: *knowledge gained.*

Evaluating *changed behavior* and *improved results*, Kirkpatrick's third and fourth levels of evaluation, is by necessity a longer-term project than the first two levels of evaluation described above. The basic legal requirement for the county DFS offices was to increase the percent of welfare recipients who were involved in some type of job activity, either training or work. The central DFS staff had the results of job involvement efforts for each county prior to the training. It was comparatively easy to evaluate the effects of the training through changes in these performance numbers after completing the five months of learning.

One word of warning about evaluation of learning is important. It is never possible to totally isolate the effects of a learning initiative like this training program on an organization. For example, with the DFS training, other factors might have influenced performance. Changes in legislation or state regulations allowing the county offices more or less freedom in doing their work could affect performance independent of the learning experience we provided. Still, improved performance is highly desirable in any organization, whether we know exactly what caused it — or not.

In this large, southern Missouri project, evaluating improved performance was somewhat more complicated than for the other areas of Kirkpatrick's evaluation model: reaction, content learning, and results. However, an organizational climate survey was completed prior to the learning workshops for county workers, and was then available for completion after the seminars were done. This means there was the possibility of a pre- and post-evaluation using *employee perspective* on the behavior of employees in the county office.

In summary, evaluation of learning initiatives is important. Efforts to evaluate all levels of learning program impact can be productive, but are never simple. People and organizations are complicated. The best one can do is to collect objective information about the impact of learning of a program and then use one's judgment to assess the effects of the learning experience based on the available data.

9. The use of *multisensory learning approaches* adds to the effectiveness of training and development. This is connected to the requirement to customize learning experiences to the Identity of the learners, discussed above. It is well established that different people learn differently; some learn better from seeing, some from hearing, and some from activity, like writing or role-playing. It is usually not possible to assess every participant's learning style before a learning program. That limitation is certainly true for a large program like that for the Division of Family Services discussed above. That generally means that using multi-sensory inputs is important in all programs, allowing the participants to learn the way they learn best. The technical approaches available today, from PowerPoint presentations to interactive computer-based training to virtual reality simulations, make the use of multisensory learning approaches very achievable.
10. The inclusion of methods for reinforcing and using learning back on the job is the ultimate test of whether the learning is being supported by organizational leadership. As discussed repeatedly, leadership support is key to making the learning effective back on the job. The support of leadership should include their involvement in the actual learning experiences while the program is in effect. Follow-up to reinforce use of the learning back on the job is also a key to determining whether or not the training and development have been worth the time and expense.

Part III of this book discusses the methods available for making learning permanent so that it can impact performance. Here, it is important to note that customizing learning to the participants' performance needs and reinforcement of that learning are crucial to performance. Learning, whether through organizationally supported education, training, or development, should be evaluated with reference to the ultimate test: *did the learning help those involved perform better than before?*

SUGGESTED ACTION STEPS FOR ORGANIZATIONAL OR TEAM LEADERS

1. Make sure that your strategy has been well designed and communicated to those required to provide or experience learning. Help make the connection between strategic accomplishment and learning initiatives.
2. Determine how important you believe learning is for achieving the organization or team strategy.
3. Identify and begin actions in those areas where your support as a leader will enhance dedication on the part of learning participants. Keep in mind that this is important not just during the training or development, but also after the structured learning experiences have occurred. Consistent reinforcement is the key.
4. Make sure that any learning facilitator involved with your organization or team has as a primary motivation that of helping people learn. Personal benefits will exist but they must be seen as legitimate only as a result of effective facilitation and learning.
5. Whether you use learning facilitators from outside or inside your organization or team, make sure they appreciate and use methods and approaches that make the learning effective.
6. Make sure that your learning initiatives have methods for evaluation on all four of Kirkpatrick's measurement levels.

END NOTES

1. Bassi, Laurie J. and Van Buren, Mark E., *State of the Industry Report,* Training and Development, January 1998.
2. Blanchard, P. Nick and Thacker, James W., *Effective Training,* Prentice Hall, 1999, p. 8ff.
3. For example: Blanchard, P. Nick and Thacker, James W., *Effective Training,* Prentice Hall, 1999; Mesiter, Jeanne C., *Corporate Quality Universities,* Irwin Professional Publishing, 1994; Brinkerhoff, Robert O. and Gill, Stephen J., *The Learning Alliance,* Josey-Bass Publishers, 1994; Robertson, Dana Gaines and Robinson, James C., *Performance Consulting,* Berrett-Koehler, 1996.
4. Drucker, Peter F., *Management Challenges for the 21st Century,* Harper Business, 1999, p. 142ff.
5. See T + D, published by the American Society for Training and Development (ASTD), September 2006, "Preparing the Workforce," p. 33ff.

6. Many excellent books on training and development, including those cited in End Note 3, are concerned with improving training and development. Perhaps no publication has been as dramatic in making the point about the limited effectiveness of learning experiences as the nationwide headlines a few years ago in *USA Today*. "Big Lesson: Billions Wasted on Job Skills Training," *USA. Today,* Wednesday, October 7, 1998, front page.
7. Robinson, Dana Gaines and Robinson, James G., *Training for Impact,* Josey-Bass Publishers, 1989, p. 109.
8. Kirkpatrick, Donald L., *Evaluating Training Programs,* Berrett-Koehler Publishers, 1998.
9. See, for example, Schorr, Lisbeth, *Common Purpose,* Anchor Books, 1997.

PART III

PERFORMANCE IMPROVEMENT: FORCES WORKING FOR AND AGAINST NEEDED CHANGE

Chapter 16

FACTORS AFFECTING PERFORMANCE IMPROVEMENT IMPACT AND STABILITY

INTRODUCTION AND LINKAGE

To this point, this book has discussed the following major topics:

- Part I discussed the powerful drive in our society for performance improvement by organizations that has led to massive expenditures of money and effort. It argued that years of experience and significant amounts of research indicate that a lot of performance improvement efforts have limited benefits to the organizations trying to improve. It identified three critical success factors for performance improvements to have a major impact on organizations and teams. The most important success factor is dedicated organizational leadership. In addition, Part I provided:
  - A definition of performance
  - A discussion of the importance of leadership providing a thorough strategic vision for the organization or team
  - A discussion of approaches for the identification of performance gaps
  - A model of planned change for identification and management of performance improvement where and when it is needed
  - A discussion of performance measurement processes that can be used in new organizations

  - A discussion of the benefits and characteristics of goal setting and its role in performance
  - Numerous cases to emphasize the main points made in Part I
- Part II began with a discussion of the importance of building a culture and structure for a learning organization. Learning new concepts and skills is the basis for high performance by individuals, teams, and organizations. In addition, Part II provided discussion of a series of performance improvement interventions, including:
  - Selection and hiring
  - Strategic planning and cascading goals
  - A model for effective teamwork and team building
  - Developing leadership and management skills
  - Programs focused on improving effectiveness and efficiency
  - Technologies and approaches for performance improvement at the individual level
  - Training, development, and education as stand-alone performance improvement interventions
  - The use of training, development, and education to support other performance improvement interventions, a deficiency in many organizational efforts at improvement

Part III, which begins with this chapter, consists of three chapters, each dealing with extremely important topics regarding improving performance in organizations and teams.

1. This chapter (Chapter 16) identifies the powerful forces requiring organizations of all types to make dramatic changes and adjustments in their strategy and operations. Also discussed is how the numerous performance improvement efforts conducted by organizations are resisted by *forces that work against change.* The forces working against change section is followed by what we know about "techniques for encouraging change." Finally, this chapter concludes with recommendations about making performance improvement a permanent part of the organization. This means preparing the organization to be receptive to and make effective use of specific performance improvement programs.
2. Chapter 17 looks at performance improvement from the leadership perspective. How can organizational leaders best link the various available performance improvement interventions into a single approach to make performance improvement a permanent part of the organization's culture and way of doing business?

3. Chapter 18 asks a question occasionally heard from CEOs and other top managers. The basic question is "Why?" Why should I do all this? What are the risks of engaging in changing the culture and organizational dynamics toward performance improvement as a permanent dedication? What are the benefits of that change?

EXTERNAL FORCES REQUIRING CHANGE BY ORGANIZATIONS

Many experts in the field of Organizational Development cite the powerful forces external to organizations that require them to change. "In the future, the only winning companies will be those that respond quickly to change. ... Organizations are never completely static. They are in continuous interaction with external forces." (1)

The external forces requiring organizations to change are numerous and fluctuate rapidly. The more obvious forces requiring change are new or expanded competition, customer demands, government regulations, changes in the economy, changes in suppliers, and changes in our sources of labor. There is no doubt that the ability of organizations to change is critical to their survival. *But change in and of itself is not sufficient.* The key is for organizations to *make the right change.*

Organizations must change in ways that help them survive and become successful. However, many organizations working to change disappear every year. According to *The New York Times,* "A record number of businesses failed in the United States last year (2005)." (2) BizMiner, a marketing research organization, says that 22.3 percent was the overall failure rate for the United States in 2005. (3) A large percentage of these organizations changed, but failed because the changes were unproductive or ineffective. Approximately 30 percent of all start-up businesses fail within a year or so as they try to make the changes required while starting a new business. Enron grew dramatically, got into highly diverse businesses including power, pulp and paper, and communications. *Fortune* named Enron "America's Most Innovative Company" for six consecutive years. It became clear later that Enron's financial condition was sustained by accounting fraud. The May Company tried for years to change its customer service approaches, and still ceased to exist when it was bought out by Macy's.

Whether to change or not to change is not the issue. How to decide what changes to make is the first decision. The second decision is how to make the changes happen in ways that enhance performance. Chapter 17 provides an overall approach to answering those two questions. The title of that chapter is accurately descriptive, "Comprehensive Performance Improvement: Actions for Leadership."

FORCES WORKING AGAINST CHANGE

There are three basic levels of resistance to change: (1) the culture, (2) the organizational structure and systems, and (3) the nature of the people in the organization.

The *cultural forces* working against change are mostly about the beliefs and habits ingrained throughout the employee population. As examples:

- "We have been very successful in the past, why should we change?"
- "We are a family here and do not want to hire people who are not part of our informal family, even if they have skills we need."

Sometimes these beliefs keep employees from seeing that change is needed, that the status quo is not desirable. Sometimes these beliefs are the source of resistance to change by those who like things "the way they used to be."

There are many *organizational structures and systems* that resist change. For example, the organization may have a hiring process that looks for people "who fit well here." That means it will keep others outside the organization. The organizational structure also identifies who the "experts" are in product quality, marketing, finance, production, and Human Resources. By definition, others should not presume to know what might need change in areas outside their responsibility. The standard operating procedures will emphasize rules and accepted ways of doing things. The organizational structure tells us who reports to whom and who is responsible for what. It also tells employees to keep their noses out of other people's areas of business, even if something is being done incorrectly. Finally, resistance to change comes from the self-serving concern about keeping one's budget or other resource allocation intact. Change that might lead to reduced power is usually resisted.

If the organization and the culture are change resistant, they will end up hiring and rewarding *people who are somewhat change resistant,* or at least slower than most people to change. Generalization is always dangerous and ignores individuals who are exceptions to the general rule. Having said that, some types of organizations and vocations encourage an above-normal percentage of people who are slower than the average person at changing. Personality profiling will usually confirm that large bureaucracies tend to attract people that like stability. Examples are institutions of higher education, large healthcare organizations, and government agencies. Stability is at a comparatively high level in these organizations.

Some types of professions and vocations also tend to attract people who like stability above change. People heavy into technology are systems oriented and often go through change systematically, rather than rapidly. Accountants spend a great deal of time working with a set of systems and procedures for managing finances, and people who like stability are sometimes attracted by this profession. The same is true of engineers and educators. These are fine professions, and we are not being critical of them. We are simply arguing that the willingness to accept change and make it happen varies by individual, and individuals tend to move toward professions that make them comfortable. People seeking stability seek stable situations, like large organizations that are slower to change and professions with lots of structure, systems, and procedures.

TECHNIQUES FOR ENCOURAGING CHANGE

There are two basic situations where organizational change should be encouraged. The first, and easiest to achieve, is change directly related to an ongoing performance improvement intervention. Thus, for example, when the organization is beginning a program to improve the style of top leadership, those trying to improve their methods of leading need to work at achieving change in how they manage and lead. Others working with them, including their subordinates, peers, and superiors, should expect, support, and encourage changes in how they lead and manage.

The second and most difficult organizational change occurs when the entire organization is required to undergo change to achieve performance improvement in many different areas: its culture, organizational structure, and at least some of its basic systems such as product offerings, customer service, and various human resources procedures. The literature uses many different terms to define broad-scale organizational change: broad-scale cultural change, organizational transformation, and strategic transformation. Whatever term is used, organizational transformation has the following characteristics. First, it tends to require many changes from the top leadership level to the bottom employee level. Second, it requires many changes horizontally across many different organizational departments and divisions: production, finance, customer service, etc. Finally, organizational transformation takes time to have major impact and become stable. The textbooks usually argue that organizational transformation takes five to fifteen years for major changes to become accepted.

Both types of changes, those targeted by a specific performance improvement intervention and full-scale organizational transformation, require the use of four basic approaches for the performance improvement to become permanent:

1. *Organizational leadership* plays a central role in supporting performance improvement.
2. The performance improvement interventions are directed toward changing some characteristics of the organization and its culture, systems, and structure.

3. The performance improvement program deals with and may *change the behavior* and supportive values and attitudes of *the people* involved.
4. The approach for deciding on and *managing the performance improvement intervention* or overall program develops broad-based support for the needed changes.

First Technique for Encouraging Performance Improvement Change: The Critical Role of Leadership

Direct experience with hundreds of organizations combined with writing from many authors in performance improvement mutually support the virtual certainty of the following principle:

The dedicated and persistent support of organizational or team leadership is often the most important factor in making performance improvement happen and maintaining progress once it is achieved. This is true whether the performance improvement intervention is shorter term, like re-engineering the marketing department, or an organizational transformation that may take many years.

Diagnosis of performance gaps and their causes is very important in performance improvement. Without an accurate diagnosis of barriers to high performance, change interventions can target the wrong issues. Without strong leadership support and involvement, diagnosis will usually lack important data about and sufficient insight into the real problems. In addition, how seriously the performance intervention is taken and whether effort is maintained after the initial enthusiasm also depend on leadership's appropriate involvement. What responsibility the leadership of organizations or teams should take to make performance improvement work is best understood through use of the Model of Planned Change (Figure 16.1), originally discussed in Chapter 3.

The role of organizational or team leaders is crucial at all stages of a performance improvement intervention: entering and contracting, diagnosis, conducting the actual intervention, and evaluation and stabilization of its results. Designing or supporting a process to deal with performance issues can strike anxiety in the hearts of even the most courageous organizational or team leader. The most common fear of leadership is included in the following reflection we have heard from many top leaders:

> "What if I am a major cause for the performance deficiencies we have?"

If possible, leadership should be involved in all stages of the model for planned change, especially if the performance improvement targets are

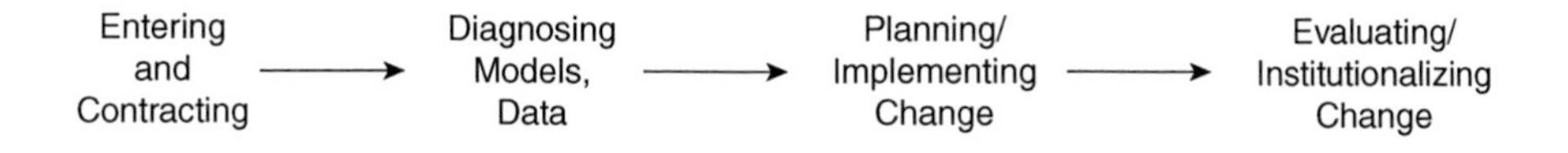

Figure 16.1 The Model of Planned Change.

key to the success of the organization. This starts with their participation in the initial discussions that usually occur during entering and contracting. See Chapter 3 for a review of this step in the process.

Willingness on the part of leaders to support the *diagnostic stage* in the process where the causes of performance deficiencies are honestly identified requires a great deal of dedication and confidence from those in leadership positions. If internal or external performance improvement consultants are also involved, trust between consultants and leadership is often a greater issue. Full candor about what is working and what is not working, and the willingness of all participants to discuss the feedback, is the key. The point is that leaders set the tone for this phase by showing whether they want real answers no matter what the data indicates.

Thus, it takes a good deal of courage for leadership to be fully candid about performance gaps and what they may have to change during a performance improvement intervention such as restructuring the top levels of the organization. But leaders of an organization must have another personal characteristic to oversee *massive organizational transformation*. That characteristic is incredible endurance for staying the course.

Diagnosis of a performance issue involves the technical procedures of data collection and interpretation of what the data says about performance. It is important for leadership to stay closely involved during feedback and discussion about what the data seem to indicate regarding performance, successes, gaps, and the causes of those gaps. This does not mean their blind acceptance of the interpretations about the data. Interpretation of data in areas of human performance is always based on judgment on the part of the interpreter, and is therefore not to be taken as "gospel." But a willingness on the part of organizational or team leaders to understand and consider the interpretation regarding the causes of performance results and the causes of performance gaps is a must if serious improvement has any chance of occurring.

In the interpretation of data, there is value for the role of performance improvement experts, especially those with no permanent responsibility in the organization. Where possible, performance experts, whether from inside or outside the organization, should be included in the process of evaluating what the data is saying. If these "experts" are not a part of the leadership team, they will have fewer emotional ties to the results of the diagnosis. A performance consultant from the Human Resources function of a large organization, for example, is one possible source of judgments about performance that may not be as influenced by personal agendas. Performance improvement experts from outside the organization are another possible source for interpretation of the data.

The second requirement for leadership during diagnosis is to be fully involved in deciding about whether a performance improvement intervention is needed — and if so, what it should be. A related following decision is the involvement of leadership in any selected performance improvement intervention. Clearly, the leadership needs to be deeply involved if the selected performance improvement intervention is strategic planning. The role of leadership in other interventions that might be selected is a matter of answering the following question:

> "What involvement should the leaders have in the performance improvement intervention in order to maximize the effort directed towards that intervention and the improvement results it achieves?"

The third stage of the Model of Planned Change, the *intervention* itself, requires varying degrees of leadership involvement. The following scenarios clarify the *options leaders have in deciding on their best role* during the intervention process.

Scenario 1. This first option requires a shorter-term, but very important role of top leaders in a performance improvement intervention. The organizational or team leaders clarify their beliefs about what performance level the team needs to accomplish. This should include what current performance is acceptable and where it appears that some improvements are needed. The leaders should clearly state their support for the work the intervention team is planning to do for performance improvement. This should include identification of any resources the leaders can provide to make the intervention happen. If the leaders are not heavily involved in the functions of the team managing the intervention, then the intervention will not be taken as seriously as when the leaders are involved.

If the leaders are directly involved in the daily functioning of the team, then it is especially important for leaders to be directly involved during the intervention. For example, if a top leader is involved in the sales team, the leader and the members of the group should work together in redesigning sales functions: creating new market-to-product connections, identifying key performance indicators to measure future selling activity, or going through sales training to learn better closing techniques.

Scenario 2. The involvement of leaders in the performance improvement efforts is longer term than in Scenario 1 above. Here, the leadership of the organization starts with the support identified in Scenario 1. However, the process for performance improvement is designed to generate a substantial amount of information and future recommendations on needed performance improvements. These recommendations for further improvements require that leadership stay involved in the intervention on a long-term basis because of the importance of the intervention to the success of the organization. The following brief case study illustrates how leadership can profitably stay with a performance improvement intervention on a long-term basis.

Organization: Division of Family Services, Boot Heel of Missouri

- Products and services:
  - Welfare cash benefits, childcare, employment training and referrals, healthcare benefits
- Markets:
  - People on the welfare rolls who were under legal pressure from Welfare Reform
- Structure:
  - Area III of the Missouri Division of Family Services
  - Headed by an area director and other central area staff
  - More than 20 southern Missouri counties formed into Area III
  - Each county delivered products and services to those eligible in their county areas
- Issues and problems:
  - One might remember that Missouri took the initiative, prodded in part by the Dean from a college of Health and Human Services, to undertake a massive training and development program for nine of the county welfare agencies responsible for making "Welfare to Work" happen. More than 250 participants were included in 40 hours of training and development spread over five months of effort. As part of the pre-work, all of the participants completed a brief organizational climate assessment of their county. This included assessment of organizational clarity, communication, decision-making, and allocation of responsibility in the county organizations.

    During the training sessions, participants were organized into small teams by county. Each county team took the data regarding their county that came from the organizational climate survey and used a problem solving process to decide how to improve the climate in their organization. Recommendations for improvement included rearranging workload, the use of new methods for communicating the frequent policy changes, and continued development of the leadership and management style of directors and supervisors. Many of these recommendations were supported by the climate survey data and were taken very seriously because they were the opinions of a large number of staff members. The area director, responsible for the entire group of participating counties, stayed closely involved throughout the many months of problem solving in the training program. Her approach was to evaluate each recommendation and, where possible, implement a large percentage of the recommendation by working with the director responsible for each county. Predictions are that this *long-term dedicated involvement* from the top executive in Area III of the Missouri Department of Family Services will pay performance dividends.

Scenario 3. This scenario requires very long-term leadership involvement in performance improvement efforts. This type of permanent leadership involvement is exemplified by the case study of Ranken Technical College.

Organization: Ranken Technical College

- Products and services:
  - Technical education
- Markets:
  - Adults
- Structure:
  - Central leadership team of about eight people
  - President and vice president with strong personalities
  - Two faculty groups: general studies and technical studies
  - Staff
  - Approximately 3000 students
- Primary issues and problems:
  - Evidence from an organizational culture survey and discussions with some influential leaders in the organization showed that power was overly centralized, with too few decisions made by people who had to implement them
  - Some major interpersonal conflict between the two top leaders and others in the organization such that performance was probably being damaged in part because of the time and focus spent on unproductive disputes

The leadership group of the college decided to go through a development process in which they learned the High Achieving Manager leadership model. The process they agreed to included taking actions to manage more like the Achieving Manager. In this process, the participating managers received feedback from at least three direct reports using survey instruments that positioned them in relationship to the High, Average, and Low Achieving Managers. They then set goals and action steps to improve their style to become more like the High Achieving Manager. This included a series of meetings with their direct reports to discuss specific steps in leadership and management improvement. The project lasted about six months.

Part of the intervention included conducting an Organizational Climate Survey, receiving data from 100 percent of the employees. The CEO of the college, having a dominant personality, consciously worked to remain less outspoken about the results and the suggested actions to improve the organizational culture. While it was clear that he wanted improvements in culture, he was not going to "dictate" changes. He wanted other leaders to take major responsibility for improvements. He was fully involved in group decisions as the team discussed improved leadership styles and moving the power of decision making further down into faculty and staff groups. He also received feedback from his own direct reports, and fully participated in working with those direct reports, as well as others, to improve his leadership style. This leader was totally involved in the intervention, but did not over-control the process.

The fourth stage of the General Model of Planned Change, *evaluation and stabilizing performance improvements,* requires the power of leadership position in making sure that a frequent deficiency in improvement efforts does not occur. That deficiency is the *dissipation of the progress* after the intervention effort begins to wind down.

There are a number of ways that leadership can be involved in "stabilizing" performance progress. One is through the power of recognition. There is plenty of evidence that many managers in the United States operate on the "exception principle of leadership"; that is, they only pay attention to problems and tend to ignore positive occurrences. This is not only poor leadership, but it is also ignorant of a basic principle of human behavior. That basic principle is "…rewarded behavior tends to be repeated."

To his credit, the CEO of Ranken Technical College, discussed above, worked hard to overcome his personality-driven tendency to ignore the positive and focus on the negative. Sincere recognition and reward for performance progress is not only good ethics, it is good motivational psychology.

Leadership should also make sure that areas of performance improvement are measured and evaluated in performance management systems. Chapter 17 discusses this more fully, but an example here helps make the point clearer. Assume that a bank has decided that "cross-selling" (i.e., one department referring clients to another department for additional sales) should be seriously emphasized. So loan officers referring to the credit card department, for example, should be recognized and rewarded. First, some brief education and training about those procedures and goals should be conducted. After that, establish a recognition and reward system for successful internal referrals. It would also be wise for the leaders to make sure that their *performance management system* includes goals for and evaluation of the referral activity. This approach would combine skills training with performance recognition and rewards, and, finally, measure of performance in the individual performance reviews.

Longer-term organizational transformation consists of two basic plans. First, an overall plan or vision for how the organization should look and operate at some point in the future, the future point being at least five years away. The second plan is a series of shorter-term interventions that are steps toward the longer-term plan. Obviously, the primary responsibility for achieving longer-term organizational transformation resides with the leadership of an organization. This starts with leadership being actively involved in creating both plans and the connection between the longer-term transformation plan and the shorter-term interventions. Chapter 17 lays out a comprehensive format for each of these plans; it also includes three organizations that have made major strides toward their transformation.

Second Technique for Encouraging Performance Improvement Change: Changing the Organization and Culture

Organizational culture is one major factor in how well an organization or team does at stabilizing performance improvements once they have occurred. The nature of the organization's culture also will impact how well they achieve organizational transformation. In fact, changing the culture is a central focus of organizational transformation. First, an organization's culture is defined as its shared beliefs, values, and behavioral norms. This results in the visible aspects of the organization such as structure, systems, methods, consistent behavioral styles, procedures, work rules, and physical trappings such as equipment, technology, and machines.

There has been increasing interest in organizational culture, a term borrowed from cultural anthropology, as having major importance for organizational change and performance.

> "There is considerable speculation and increasing research suggesting that organization culture can improve an organization's ability to implement new business strategies as well as to achieve high levels of performance." (1)

Which dimensions of organizational culture most influence the abilities to adopt new strategic directions and achieve high levels of performance are matters of controversy. Experience, however, suggests strongly that the following dimensions are most significant:

- Goal-setting, measurement, and results focus
- Open communication and a willingness to discuss even the most difficult topics
- Valuing leadership that is oriented toward both results and people development, and strong support for at least some of the current organizational or team leaders

Without a strong cultural orientation toward goals, measurement, and results, no one really knows the status of performance and where the gaps exist. Without the willingness to openly discuss even highly sensitive performance issues, difficult topics and issues are not confronted, and things go on as usual. Without appreciation for a leadership style that values both results and people, the High Achieving Manager discussed in Chapter 9, either people are de-emphasized and their motivation declines, or everything is comfortable and nothing improves. A brief case study of a medium-sized printing company clarifies this point:

- Products and services:
  - Printing with specialization in corporate annual reports
- Structure:
  - New CEO after many years of leadership from a now semi-retired owner
  - Vice president for sales and CFS
  - Divided into sales group and production group
- Issues and problems:
  - There were major conflicts between production and selling over issues of scheduling work and delivery times, a classic production/sales dispute seen in many organizations
  - There had been prior attempts to solve the work scheduling problem; but after some apparent agreement about actions to improve performance in this critical area, little happened
  - At least some of the sales and product employees did not believe that the leadership group would make them do what was needed to reduce scheduling conflicts and production errors and improve on-time delivery to customers

The sales and production leaders, prompted by their leadership team, experienced a team problem-solving process that led to further recommendations on improving the work scheduling and delivery problems. The recommendations were never really installed as people got back to their daily routines and forgot their agreements. The leadership group was unable, or unwilling, to require that the sales and production managers put the recommendations into effect. This organization is no longer in business after a fire that destroyed its buildings.

This case involves a culture that was short on holding people accountable for solving problems and keeping their commitments to make performance improvements work. Human beings behave in the direction of their values, beliefs, and learned habits. However, those values, beliefs, and learned habits are influenced by the organizational culture in which they operate. When the culture does not emphasize accountability, even the most responsible employee finds it difficult to remain motivated. It is difficult to work against a culture that lets others ignore accountability. When the culture is strong in recognizing and rewarding desired values, beliefs, and habits, then employees either become "acculturated" in those directions, or withdraw and perhaps even leave the organization. Ironically, a culture that is strongly dedicated to certain beliefs, values, and ways of doing things can often produce excellent performance results. But strong cultures are often slow to change because of satisfaction with current performance. The ability to change a high-performance organization toward a dramatically new direction is difficult because current patterns are so strongly reinforced. Still, when the case can be made that continued high performance requires some changes in the organization, high performers will become motivated to make needed changes if they buy in to the supporting arguments.

The *structure of the organization or team* also influences the ability to put performance improvements into effect. The above case study citing the printing company was partly an issue of leadership and culture, but it was also partly a structural problem. The separation between production and sales led to different agendas and constant conflict between the departments. There are common structural arrangements that create problems affecting initiatives to improve performance. These arrangements include:

- Divisions between sales and production with no linkage or central authority resolving disputes
- Divisions between marketing and sales with those selling directly to the clients often having very different perspectives about what is needed in marketing strategies from those making marketing decisions
- Organizations with field-based personnel spread out across wide geographic areas having little support from or responsibility to either a regional or central authority

In these, as well as other structural scenarios, the performance problems usually come from a lack of a central authority having dedication and persistence to improve performance and keep it at a high level. When people are separated by functional responsibility and reporting relationships or by geographic separation, someone must be responsible and empowered to overcome those separations.

The *location of the ultimate decision-making power* is also a key in making performance improvements permanent. If those ultimately responsible for performance in an organization or team are outside that group, then performance improvements always must go through "higher-level" approvals, and may well be vetoed by executives. This is the frustration of working in a large organizational structure where a relatively independent team is responsible for improving its results but often does not have the final authority to make decisions that can produce the desired results.

Structure is designed by people and can be changed by organizational leadership. The requirement is to use the guidelines of effective organizational and team performance when restructuring. See Chapter 12 for a more detailed discussion.

Third Technique for Encouraging Performance Improvement Change: Changing the Characteristics of the People in the Organization

When the organization culture emphasizes results as well as commitment to people development, it will tend to attract people who are willing to take on challenges, grow, and learn. When the organization culture has emphasized stability and routine, it will tend to attract and keep people who are much more comfortable with stability and often less willing to change as needed. The Missouri Division of Family Services case, discussed previously in this chapter, is a classic example of the impact of group personality on the speed and ability to change for improved performance. Remember that the county welfare offices are newly responsible for labor exchange, finding jobs for welfare recipients as they move from welfare to work. This is very different from the determination of welfare eligibility and granting of benefits that the welfare workers have done for many decades. But there are major struggles in making those changes. The initiative with the Division of Family Services included a personality profile of all 250 participants for personal development. Out of four dimensions measured by this particular profile, the dimension measuring a tendency toward stability and steadiness was *the primary dimension for close to 70 percent of the* 250 *welfare workers*. It is not that they could not change to deliver the new services they were required to provide. Some of them were working to change what they did at work. But, in general, they changed slowly. Others chose to leave the organization out of frustration over "unreasonable changes."

When major changes are required in what people do at work and how they do it, one of two things will usually result. A number of those needing to change how they perform will do so. Some personalities change more easily, others with greater difficulty. Others will choose to leave the organization or will be asked to leave. At that point, the organization has the opportunity to hire those who more closely fit the transforming organization.

Fourth Technique for Encouraging Performance Improvement Change: Conducting Effective Performance Improvement Interventions

1. It is imperative that any performance improvement initiative maximizes the opportunity for those in the organization or team to diagnose their own performance and build their own action plans for improvement. During diagnosis prior to the actual performance improvement intervention, the performance experts should have clearly identified performance gaps, causes, and perhaps even some potential solutions. However, the work at improving performance must include activity where those involved in the organization/team come to see the same issues identified in diagnosis, and then they refine and add to that diagnosis. Those who are to make performance improvements happen must also be involved in designing changes and actions for improvement. When outside consultants have undue influence in diagnosing situations and developing performance improvement recommendations, the program will be resisted. This is called the NIH syndrome — "not invented here."
2. The performance improvement intervention must *set clear goals for action* with dates and responsibilities. This is easily done with strategic planning because setting objectives is a final step. But areas such as improving management style, decreasing defects in the product, or building better customer service may require effort in setting goals and deadlines. Still, the efforts are worth it.
3. *Performance improvements must be measured and recognized* when they occur. This is another place where the dedication of leadership at keeping performance improvements on center stage is essential.
4. The performance improvement initiative *must include those in decision-making authority* over the organization or team needing development. This is often difficult to do in a larger organization where plants or sales teams, for example, have a good deal of freedom and may want to work on their own performance, but do so independently of central executives. However, the leaders of these independent groups are well advised to keep higher-level

executives aware of what they are doing, even if those executives are not directly involved in the actual initiative.

5. Perhaps most importantly, leadership must share the overall direction of the organization as the changes are occurring. This should be done frequently and openly. The now-popular "town hall" approach is as good a process as any for keeping people in the know about what is happening to the organization in which they work. Town halls are open to questions from the floor regarding both long-term direction and immediate interventions and programs.

SUMMARY AND CONCLUSION

The four factors identified above that affect the ability to evaluate and stabilize performance improvement are seldom all in perfect shape in any actual situation. However, improvement in stabilizing performance improvement can also be achieved. For example:

- Leaders can be taught the importance of their role in making improvements permanent and can learn how to do it.
- Organizational cultures that do not emphasize goals, standards, measurement, and performance results can be changed through installation of goal-setting systems and performance measurement.
- People who have personalities and learned habits that work against changing to improve performance can and do change. Even those whose behavior is highly habitized can learn new habits, although this, as well as culture change, takes time.
- There are many ways to design performance improvement interventions so as to maximize participant involvement in diagnosis of the issues and action planning for improvements.

It all takes dedication and persistence.

SUGGESTED ACTION STEPS FOR ORGANIZATIONAL OR TEAM LEADERS

1. Identify the areas where performance improvement efforts need to occur. What areas of your organization or team are essential for you to have high-level performance to be successful?
2. Do you and your managers have sufficient goals and measurement to know if performance is sufficiently high? If so, do you have significant performance gaps? What are the causes of those gaps?

3. If you do not have sufficient goals, standards, and measures to know the performance level in essential areas of your organization, start developing those measures. Use concepts of goals, KPIs, and standards discussed here and in other referenced publications. Once you have a measure of performance success and gaps in important areas of your organization or team, go back to Action Step 2 above, and then proceed with the list below.
4. If significant performance gaps exist in your organization or team in areas that are essential, what performance interventions are available to effectively overcome those gaps? Stay with this process of decision making until the intervention is identified and put into effect. Make sure you or someone in your organization who is determined to see improvement stays with the project to initiate the performance improvement intervention as it unfolds.
5. Stay involved with the intervention sufficiently to learn what is being done and to show your commitment to that process. If there are knowledge, skills, attitudes, and motivation needed to improve your own performance, be fully involved in the intervention. Remember that your involvement will be a major factor in how others in your group see the importance of the initiative.
6. Evaluate the intervention. What worked and what did not work? What needs to be kept and reinforced? How can you best do that? Look at this chapter and Chapter 17 to get some ideas about reinforcing desirable performance improvement change.
7. If you are the leader of an independent team within a larger organization, find ways to keep higher-level executives aware of what you are doing for performance improvement.

END NOTES

1. Harvey, Don and Brown, Donald R., *An Experimental Approach to Organization Development*, sixth edition, Prentice Hall, 2001, p. 3.
2. *The New York Times,* September 28, 2006.
3. BizMiner, U.S. Marketing Research Profile, pp. 1–6, see www.bizminer.com/2006-Profiles/samples/USMR.asp?sic
4. Cummings, Thomas G. and Worley, Christopher G., *Organizational Development and Change,* South-Western College Publishing, 1993, pp. 480, 481.

Chapter 17

COMPREHENSIVE PERFORMANCE IMPROVEMENT: ACTIONS FOR LEADERSHIP

Does my organization need major transformation?
If so, how is it done successfully?

INTRODUCTION

The first chapter of this book discussed the negative consequences of letting the wrong reasons, such as current popularity of performance improvement initiatives, motivate improvement efforts. This usually results in little or no diagnosis of real issues, and therefore ineffective change programs. Stated more directly, the initial key stage in performance improvement is *diagnosis*. Conducting programs for reasons other than clearly identified performance gaps and improvement needs is wasteful and frustrating. It takes organizations, teams, and people in the wrong direction.

One of the most difficult questions to answer during diagnosis concerns the scope of change that is needed. The question is: how many areas of the organization need improvement? The Organizational Success Model introduced in an early chapter divides success into two basic dimensions: (1) clearly defined strategy and (2) effective and efficient operations. Here are the kinds of questions organizational leaders need to ask in deciding the scope and nature of the performance improvements they need to make.

Strategic Questions

1. Do our products and services provide us with a strategic advantage?
2. How long before our products and services become outdated or need major changes, updating, or additions?
3. What are the major strengths and weaknesses of our products and services?
4. What is our reputation with our primary clients? Do we hold a favorable position with them? What should we do to add to our value with our clients?
5. What new clients, demographically or geographically, should we pursue? What will it take to add major markets to our current list of clients?
6. How effective is our marketing and selling? Given whatever key performance indicators (KPIs) we have about marketing and selling, how effective and efficient are these areas?
7. Is our customer service a major factor in how successful we are with keeping or adding clients? If so, what areas in customer service need major improvement?

Operations Questions

1. In which areas of my organization are effectiveness critical to our success with clients? (Effectiveness is having and achieving defined major goals.)
2. How do I rate the effectiveness in those areas of my organization listed in Question 1, just above?
   Example: 1 — Needs Help 5 — OK 10 — Great
3. What efficiency measures do we have? Are there areas where we should measure efficiency and do not do that now?
4. With the data we have, in what areas does our efficiency need major improvement?

The answers to these and related questions by organizational leadership can help the leader formulate his or her own answer to the following overall questions:

- Does my organization need changes *in one or a few targeted areas* to be as successful as we should be during the next ten years?

or

- Does my organization need changes in a significant number of areas? Are we in need of organizational transformation?

The remainder of this chapter is primarily oriented toward situations in which the organization needs major changes in a number of areas, resulting in organizational transformation. If the work to be done as a result of the leader's assessment will take some years to accomplish, and if it will involve two or more major areas or functions of the organization, then some degree of transformation is needed.

Sometimes, organizational or team leaders decide to develop their organization in a number of ways, although the organization is by all measures currently successful. This is often the result of a "sense" leaders have that the organization/team is not performing at top level, although data supporting that view is minimal. Here, the performance vision and expectations of the organization or team leaders are paramount, rather than identified areas of performance deficiency, particularly if data is minimal. The following phrase is often heard: "We are good at what we do. We want to be even better." The situation where an organization is currently successful and wants to increase its level of successful performance is one in which performance improvements are *most* apt to work. They have time to adjust and make changes.

Where performance improvement efforts are least apt to work occurs when the organization is on the brink of disaster, perhaps close to closing its doors. In this case, there is less time for them to fix things. Also, most often a lot needs fixing, which is why they are in trouble.

When organizational leaders want performance improvement and have time to make it work, they can link together a number of basic interventions. If they follow the guidelines discussed in Chapter 16, techniques for encouraging performance improvement change, then they can have success like that of the three companies profiled in the final chapter.

LINKING THE BASIC ELEMENTS OF HIGH-LEVEL PERFORMANCE

There are a series of basic requirements to good organizational/team performance. These were discussed throughout this book, using a one-at-a-time-intervention approach. Working on one area of performance improvement at a time is what most leaders choose. *Few leaders have the ambition to take on a total campaign for performance improvement, which is the centerpiece of organizational transformation.* When leaders do choose to construct a longer-term campaign for performance improvement and organizational transformation, this is what is required:

1. *The first requirement for organizational transformation toward high performance is a clearly stated strategic direction* with strategic

objectives or goals. The need for a strategic plan and immediate strategic objectives is essential for a total organization such as a large corporation, a college, or a small business. This basic requirement is met through a strategic plan as discussed in Chapter 6. Often, a strategic plan is also essential for an independent team, such as a large marketing or sales group with a good deal of discretion in how they approach their markets and which products they choose to offer to those markets. In any case, an organization, whether a corporation or a department, needs to have plans and direction to follow.

A team that has little discretion in what products or services it provides, such as production or customer service groups, would find strategic planning for its group of little use. But these groups still need to have direction for effective performance. This is where team goal setting, with goals cascading out of the strategy established higher in the organizational structure, plays a critical role. Whatever the source of direction, strategic objectives or cascading goals from a higher level, direction is a must for effective performance. This does not mean that a department should not have some freedom in deciding how to achieve its part of the organization's overall strategy; it means that the departments need to know their part in that strategy.

2. The second requirement for organizational transformation toward high performance is the *development of productive leadership style supported by management skills.* The research from the Achieving Manager project discussed in Chapter 9 indicates that more than 50 percent of all managers in the United States have serious flaws in how they manage. This is not surprising, as many people become managers with little guidance except what they have seen their own managers do in the past. Ineffective leadership and management styles get passed on. Our own experience and review of the relevant literature indicates that the High Achieving Manager is the best model for managers to learn and imitate. However, the ultimate guideline should be for managers to find ways to both work to achieve high performance goals and to develop and support their employees. Good leadership and management is not a matter of choosing between people and performance as many managerial models imply. Rather, high performance requires finding ways to achieve both effective organizational performance and effective individual performance.
3. The third requirement for effective high-level performance is to *establish and maintain techno-structural relationships* that permit effective coordination between functions and business processes

required for achieving strategic and department/team goals. Chapters 4 and 12 discussed "restructuring" an organization. There is no best structure for all or most organizations. However, it is important to recognize that any organizational or team structure should be based on the following guidelines:

a. Reporting relationships and the division of work into groups should enhance workflow in areas such as developing and producing products and services, delivering customer service, and enhancing sales and marketing results. Unfortunately, as demonstrated with a number of case studies throughout this book, many structural relationships retard these processes.
b. The organizational structure should include the ability to respond quickly to new opportunities or challenges. The term most often used in the literature is "responsiveness" or "flexibility."
c. The guiding principles for creating or redesigning an organizational structure should be to maximize effectiveness (goal achievement) and efficiency (number of inputs necessary to achieve the desired outputs).

Restructuring is an effort at performance improvement that is overused, partly because many leaders believe they understand it and know how it should be done. However, it is often done in ways that produce frustration, bad morale, and unexpected consequences. The best approach is to use people familiar with the basic business processes in an organization to recommend the restructuring (product design and delivery, sales and marketing, customer service, etc.).

4. The fourth requirement for the achievement of organizational transformation that encourages effective performance is to *know the core competencies of the key positions in the organization/team*. An overview of the meaning and role of core competencies in performance was provided in Chapter 11. As discussed, leaders interested in this performance improvement area have many sources of information available. (1) The point is that whatever method is used to define competencies, it is important to have a clear definition of underlying characteristics of an employee that are directly related to job performance. (2) This is a major flaw in most hiring processes. Not knowing what competencies are required for high levels of performance in a crucial job seriously hampers people who manage persons in that job. Organizations often lack clear and specific definitions of the competencies for performance success required for their most important jobs. Methods for identifying core competencies for various positions are covered adequately by books listed in the End Notes at the conclusion of this chapter.

5. The fifth requirement for effective performance is a productive *selection or hiring process* (see Chapter 8). Remember that sometimes changing an organization's culture toward a high performing organization requires some changes in personnel at a number of levels. For hiring to be productive in this context means at least three things:
   a. Selection or hiring must take full and specific account of the strategic direction of the organization/team. For example, when an organization makes the strategic decision to enhance competitive advantage by aggressively creating new products and services to meet market needs, people with research and development experience in the new products are needed. In addition, people with experience in selling, servicing, or producing the new products may need to be hired. This is related to the principle that a new specific strategy will impact all of the ongoing processes of the organization.
   b. Selection or hiring should measure applicant possession of the core competencies connected to the positions being filled. This is true for job applicants from both outside and inside the organization.
   c. As discussed in Chapter 8, selection or hiring should make use of all available information on the position and on the applicants. A number of people, including the person managing the position being filled, should be heavily involved in interviewing, data review, and the final candidate selection.
6. The sixth requirement for effective organizational performance and organizational transformation is to construct *programs and methods for ongoing development of the knowledge, skills, and motivation of the permanent teams and individual team members*. Those responsible for marketing, for example, must be tasked not only with meeting the basic goals of that function, but also with meeting team development goals and individual team member performance and development goals. Failing that, knowledge and skills become rapidly outdated or rusty, motivation suffers, and attitudes and morale decline. Most teams and their members want to feel competent, be competitive, and do a good job. The leadership team in the organization should also be responsible for ongoing development of their own knowledge, as is equally true for supervisors, customer service providers, and production teams. The efforts at team development by leadership will be the model followed by other teams.

   This is one of the most complex requirements for transforming an organization. This brief summary may be helpful. The learning for an intact or permanent team should focus on the following related, but different, areas of knowledge and skills:

a. The knowledge and skills related to the specific function of their department and their own professional role: marketing, production, information technology, etc.
b. The knowledge and especially the skills required for being effective team members: what an effective team is, how to help their team be good at teamwork, their own strengths and challenges in working with teams, team problem-solving processes and skills.

7. The seventh requirement for effective organizational performance is the existence of a *system of performance management that on a daily basis reinforces the first five requirements listed above.* This is where the dedication of even the most committed leadership will often falter. As often as not, the resistance of leaders to a performance management system comes from their unwillingness to provide systematic feedback to their own direct reports. It is critically important that all personnel receive constant reinforcement of the requirement for effective individual and organizational performance, recognition of high performance, and recommendations for improvement. This can be difficult because of the pressure of daily business. But performance management is at least equally important as any part of daily business. Specifically, the performance management system should:
   a. Contain a section for the identification and measurement of the employee's responsibilities for the strategic objectives as well as the operational goals and activities connected to his or her position.
   b. Contain a section for identification and evaluation of the employee's responsibilities for working with others in departments or teams separated from his or her own by the techno-structural design of the organization. That is, part of the responsibilities of the employee (whatever their level of authority and responsibility) is to find ways not to let division into organizational units stop him or her from getting his or her work done. They must complete the various business processes in which he or she is involved. For example, if someone in contact with the customer is asked for a delivery date by that customer, contacting production to get the information and relaying it to the client is a responsibility.

   It is often surprising how, even in small organizations, separate silos of activity and responsibility created by the organizational structure and personality conflicts become a barrier to effective output. This, of course, is essentially what the quality movement of a few years ago was all about. While a good techno-structural design will be based on facilitating business processes rather than retarding them, that is never enough. *People make systems and structure work or not work.* Attention to the daily performance of all employees is required. Leadership of the organization/team needs to empower their members to have access to others as required to get the work done. The leaders and managers should then make sure that empowerment is used.

c. The performance management system must reinforce attention to the core competencies required for performance of the various jobs. The importance of this comes from a subtlety about core competencies that is often forgotten. That is that they are "ideals" that are only infrequently totally met. For example, if it becomes clear that a branch manager needs to show a great deal of patience and emotional maturity with his customers and direct reports because of the unusual stress in the branch, then that behavioral trait should be recognized in the feedback to branch managers. Specific examples of adequate patience in difficult situations or lack of it should be used in performance feedback. In a simpler situation, if the position requires a good deal of technical knowledge in the use of spreadsheets for data flow, then the performance management system should include attention to that skill as well as plans for developing it in employees when it is lacking.
d. The performance management system should also reinforce the importance of a person's attention to selection or hiring if they have a role in that process. If the person being "managed and reviewed" has a part in hiring, then goals and feedback regarding their effectiveness in that process should be set and used. If the person has a part in orientation of new employees, then goals and KPIs regarding that responsibility should be set, and feedback on performance in those areas should be provided in an ongoing way. The key point here is the following:

  "Organizations get what they pay attention to and measure."

e. The performance management system should also measure the success of efforts at building the performance of permanent teams. The leader of the team should set goals and get feedback regarding his or her own efforts at developing the group for which he or she has responsibility. One of the questions that arises has to do with the CEO getting feedback when he has no boss except perhaps a Board of Directors only occasionally seen. The answer is simple. Who better to ask whether the performance of the team is being developed than the team members themselves? Top managers should provide feedback to the CEO about how he does at managing them.

8. The eighth requirement for effective organizational performance is the *need for the organization to create a learning organization*. This means two things. First, the organization develops a set of systems for identifying, storing, and updating knowledge and information it wants its employees to have, know, and use. Second, the organization must develop and use methods and systems for delivering education, training, and development aimed at enhancing performance. Requirement 6 focused on training specific to intact teams and their team members. The focus here is on learning for all persons about all subjects of importance to the organization and people within that organization.

   In larger organizations, this is usually a major focus of the Human Resources department or a related training department. The key is to avoid what has been called "training for activity" and strive toward "training for impact." This is a brief reminder of the importance of performance diagnosis as the basis for performance improvement efforts, whether training and development or any other performance intervention.

9. The ninth requirement for effective organizational performance is the *development of knowledge and skills* for one of the most difficult managerial functions: *coaching, counseling, and developing employees*. This set of abilities is best developed as the key component of the performance management system, with managers trained to use coaching and counseling. As stated previously, managers should also be held accountable for developing their direct reports. In addition, their direct reports and all the employees should be educated to understand how to make the best use of coaching and counseling they receive.

   Those who coach skills or counsel in areas of attitude, motivation, and the like need three things. The first is the commitment to the importance of this role on the part of managers and more senior employees who can help with the development of newer employees. The second is the comprehension of a series of concepts that help define the roles of the coach, counselor, or mentor. The third and perhaps most challenging need for effective coaching/counseling is the motivation and skills to do it. Somewhat surprisingly, many managers have never developed the ability to sit face-to-face with an employee and identify and discuss that employee's strengths and opportunities for improvement. This ability is, of course, much the same as what is needed in the evaluation steps in performance management. The training and education of those providing coaching or counseling can be understood in the context of the performance management system, where performance dimensions are identified to both the employee and the coach or counselor. It is always easier to discuss performance strengths and improvement needs when both parties have earlier agreed to the performance dimensions. A simple case will help clarify this point.

Organization: A Consumer Products Company

- Products and services:
  - Food
- Structure:
  - Large field sales force
- Problem and issues:
  - The business situation and the leadership style of the executive in charge of field sales and operations caused great difficulty for the CEO and the owner of the company. These two decision makers were confronted with the following dilemmas pertaining to the field operations executive:
    - The business was undergoing rapid expansion focused primarily on entry into geographic markets where it had not been previously. The business was very successful financially and promised to be even more successful if the expansion into new markets was done well.
    - The field operations executive had decades of experience that made him invaluable in the projected entry into new markets.
    - The field operations executive had, however, developed a style of leadership that was seen as totally negative and overly controlling with his direct reports. He managed by identifying problems and going after and personally attacking people who he believed caused those problems. As one of his direct reports put it, "The business is growing rapidly, we are succeeding in every way, and he sees everything as a terrible crisis."

The CEO, supported by the owner, decided that the best course was to try to change the leadership style of the field operations executive rather than lose him or continue to let him destroy morale. Turnover in field operations was higher than necessary, and fear among field staff was at a high pitch.

The CEO used an approach of constant coaching and counseling, combined with outside leadership development programs aimed at changing the style of the operations executive. The CEO made it clear to all involved, including the executive's direct reports, that personal attacks would not be tolerated. He also used two other methods that substantially improved the situation. First, he commissioned an outside study of the morale of the field operations staff. Responses were confidential, and the goal was for the survey to make clear to all involved the extent of the bad morale and the specific issues involved. While the field operations executive did not totally accept the survey showing that morale was a major issue nor his part in causing that low morale, he did begin to recognize that there was a problem that probably affected performance.

Second, the CEO moved the organization toward clear goal setting, especially in field operations. Sales goals, delivery goals, and goals pertaining to the amount of product to be left at the customers' sites were clearly established. This tended to increase the objectivity of the discussions between the field executive and his direct reports. Specific goals can take performance evaluation from vague or personal comments to somewhat more specific and measurable feedback.

The most recent description from the CEO in this coaching and counseling situation is that the field operations executive has significantly improved the objectivity of his feedback, has reduced the personal attacks, and has occasionally found the ability to provide sincere compliments for a job well done. While all is not perfect, and resentment from past attacks remains, performance remains high in the organization and morale has measurably improved.

CONCLUSION AND SUMMARY

Linking the basic elements of high-level performance into a comprehensive performance improvement effort takes enormous dedication and patience on the part of all involved. Still, it can be done successfully. Those interested in reviewing actual cases of performance improvement efforts close to comprehensive in scope can review discussions throughout this book of Scottsdale Securities, Inc. (Scottrade Inc.) and Hager Hinge.

SUGGESTED ACTION STEPS FOR ORGANIZATIONAL LEADERS

1. Review the nine requirements for effective performance and determine where you believe your organization or team meets the requirements and where it does not. Be sure to involve others in your group in the process of assessing your organization or team.
2. Determine where the deficiencies are hurting your group significantly in achieving your vision of performance and the achievement of key goals.

3. In areas where a significant deterrent to performance is occurring because of a deficiency in that requirement for effective performance, initiate discussion with your leadership team about needed actions.

END NOTES

1. Dubois, David D., *Competency-Based Performance Improvement,* HRD Press, 1993.
2. Dubois, David D., *Competency-Based Performance Improvement,* HRD Press, 1993, p. 9.
3. Robinson, Dana Gaines and Robinson, James C., *Training for Impact,* Josey-Bass Publishers, 1989.

Chapter 18

PERFORMANCE IMPROVEMENT EFFORTS: TRADE-OFFS FOR LEADERS

INTRODUCTION

Organizational or team leaders thinking about performance improvement face a profound dilemma reflected by the following question. Are the benefits of performance improvement worth the costs of those efforts? Over the years, hundreds of leaders have been heard to express the following reasons, or variations thereof, for not engaging in performance improvement:

1. "We are doing well, or at least OK now. Why mess with success? Are the benefits of performance improvement worth the effort? How can I be sure that the cost, frustrations, and energy to be put into performance improvement initiatives are going to be worth it? The benefits of performance improvement efforts seem unpredictable. Why pay the costs for an unpredictable result?"
2. "How much performance improvement do we need? How much can we take on, and when? Do we really need to make a major overhaul? Performance is difficult to define. How can we know if initiatives will work? We do not have time to work on improvement. We have to just do our work."

There are no universally valid answers to these questions. But in today's business climate, they are questions all organizational leaders must answer for their organization. The following three cases are real-world examples of organizations that decided to accomplish major organizational transformations.

Scottrade Inc.

Ten years ago, our firm began working with Scottrade Inc. Scottrade was already beginning to be successful, but its leadership decided to conduct the following performance improvement interventions:

1. Scottrade began strategic planning almost a decade ago and has conducted a strategic planning update in each succeeding year. It has worked hard to make sure that a high percentage, annually at least 50 percent, of all strategic objectives are accomplished.
2. Scottrade has spent millions of dollars during the past decade or so on developing its managers, supervisors, and team leads.
3. Scottrade recently created a Scottrade University with many different training and development programs included. Its approach is the use of a "development map" that charts the career development of the individuals experiencing education and training.
4. About a decade ago, Scottrade installed a performance development system conducted by all departments and at all organizational levels.
5. Scottrade reorganized the top level of the organization to provide closer coordination between the top executives.
6. Scottrade made extensive use of teamwork for decision making in areas such as achieving its strategic objectives.

What are the *key performance indicators* (KPIs) for this successful company?

1. Scottrade has gone from approximately 60 branches when it started its transformation about a decade ago to a system that is approaching 300 branches.
2. Scottrade's number of employees has increased approximately 400 percent.
3. Ten years ago, Scottrade's sole product was providing stock trading opportunities. Now it has a full line of equities, mutual funds, and other investment products.
4. Scottrade has an Asia-Pacific desk that has hundreds of thousands of accounts from Asians in both the United States and other parts of the world.
5. Scottrade has won the J.D. Powers award six years in a row for providing the best customer service for the online discount trading industry.
6. Scottrade has advanced beyond most companies in having a business-driven IT program. This program is driven using a committee to determine IT objectives, a committee that consists primarily of business executives from this firm.

Landshire Inc.

We first met the leadership of Landshire 15 years ago. The leadership in this company also made major organizational changes, including:

1. Landshire started with a management development program, and has continued each year to engage in management training primarily using outside sources.
2. Approximately four years ago, Landshire undertook strategic planning for the first time. Since then, it has updated its strategic plan each year. It has accomplished about 50 percent of its strategic objectives each year, including the addition of a large new market outlet for its product line.
3. Landshire installed a performance development system (PDS). Route businesses like this company work at a faster pace than some other organizations. The company decided to customize the PDS to that rapid pace by making it shorter, taking less time. However, it makes use of its PDS to formalize its goal focus, particularly for its sales force.
4. Landshire developed a full Human Resources department and significantly improved its hiring process.
5. Landshire is primarily a sales organization by almost any criteria. A large number of its employees, for example, work in field sales or a related function. Recognizing the importance of selling to its success, this organization bought and experienced a lengthy sales training and development program for all of its primary sales managers.

What are the *key performance indicators* (KPIs) for this successful company?

1. Fifteen years ago, Landshire had delivery routes in eight states. Today, its delivery routes are in eighteen states. At the earlier time it had 45 routes; it now has 90 routes.
2. Landshire has added a second major market outlet and distribution system that is nationwide. This distribution channel, which it calls "food service distribution," is done through vending company suppliers and large grocery chains. This new division accounts for a significant part of the company's business.
3. The number of Landshire employees has increased approximately 300 percent.

Alliance Credit Union

Both Scottrade and Landshire are privately owned companies. Alliance Credit Union is legally owned by its members — that is, anyone who has an account with them. The Board of Directors for Alliance consists of selected members. Alliance, originally known as Emerson Credit Union, was chartered in 1948 by a small group of Emerson Electric employees to provide financial services to members. Alliance in the 1980s diversified its primary markets to include selected employee groups (SEGs). These SEGs gave them a kind of guaranteed set of clients.

During the 1970s and 1980s, Alliance experienced major difficulties that threatened its very existence. The financial industry, including credit unions, was experiencing governmental deregulation. As deregulation occurred, two primary things occurred in the financial services industry. First, organizations providing financial services became more alike in that they provided more of the same services as other providers. Thus, banks and credit unions began to provide largely the same products. A second thing happened, largely as a consequence of increasing product similarity between providers of those products. Competition between financial service providers dramatically increased in all markets, including the selected employee groups working in organizations. Most recently, this increase in the number of financial services providers available to all of us in the United States has increased again with the advent of online banking.

The top managers and Board of Directors of Alliance Credit Union found that these trends and a few other internal developments had quickly put them in financial jeopardy. The organization needed to go through a major transformation to survive and prosper. The major performance improvement programs that the credit union has undertaken in recent years include:

1. Alliance began its journey by participating in 15 weeks of leadership and management training and development. All of its top managers were included in these sessions.
2. Sometime after the management training, Alliance conducted its first strategic planning program. This strategic plan identified objectives that were directed toward major changes in how Alliance does business.
3. In pursuit of its strategic plan, Alliance added two top managers. For the first time, the company had full-time managers for Human Resources and for Marketing. The organizational structure has expanded to include these top managers and additional responsibilities for other levels of leadership.
4. Alliance contracted with an outside supplier to lead it through a major culture change. The culture change focused primarily on good customer service and appropriate cross-selling of products where the customer's need is apparent. The culture change is measured by an organizational culture survey conducted before the culture change started and redone periodically to measure change.

5. Alliance has added personality profiling and other forms of assessment to its hiring process.
6. Alliance's Board of Directors has adopted the staff recommendation to initiate a Strategic Facilities Plan. The plan is for the credit union to add two new branches in population growth areas in their city.

What are the primary *key performance indicators* (KPIs) for this credit union?

	12/31/1991	**6/30/2006**
■ Assets	$22.5 Million	$119.5 Million
■ Deposits	$22.3 Million	$106.7 Million
■ Number of full-time equiv. employees	19	56
■ Number of branches	1	4

- Alliance currently is the largest provider of small business loans in its state, holding a portfolio of over $10 million in small business loans.

Alliance Credit Union has a lot to complete in its organizational transformation. Its efforts at culture change target the goal of building a competitive advantage over other financial services suppliers. The addition of two or more new branches will also require years to get the new facilities to top-level performance. The top managers and the Board of Directors are working together to make all of this happen.

LEADERSHIP'S BIGGEST PROBLEM

Leaders and managers of organizations often themselves resist leading major change for performance improvement. Sometimes they experience resistance from people in their organization. This can be subtle resistance or perhaps a lack of support for improvement suggestions or efforts. But however aggressive the resistance to performance improvement, the leadership of organizations should be aware of a major organizational reality: in the long haul, organizations either work to improve or they decline. Many insightful CEOs we have worked with have come to the following conclusion: "We either grow or we die!" Growth can mean more clients and assets, and it can mean doing what they do more efficiently or effectively. The leadership of the three organizations identified above have bought into their own versions of "grow."

This confusion and ambiguity that many organizational leaders feel about performance improvement efforts is the flip side of the "powerful felt need" for performance improvements discussed in Chapter 1. The reader will recall that the enormous demand for performance improvement has resulted from intense competition, changes in technology and the desire to make use of it, wanting to feel competent, and similar socioeconomic trends. Despite these pressures for performance improvements, leaders of some organizations needing improvements remain inactive or make half-hearted attempts. They are facing a dilemma. Competition and other forces indicate that working on performance is essential; however, they do not get it done.

The key to resolving this dilemma for leaders is to get specifics about performance and then decide if improvements are needed and are worth the costs in their organization. One of the most thorough methods for assessing performance for any organization is a strategic vision developed by the top leadership. During creation of the plan, diagnosis of performance gaps can occur and decisions can be made about whether improvement needs to be initiated.

Ideally, performance improvement should be an ongoing effort, with all those in an organization working to improve themselves and others for whom they are responsible all the time. But realistically, daily operations take time, energy, and effort. Sometimes, "just getting the work done" seems to be all people in the organization have time and energy to do. However, the competition is working to get ahead, products and services become outdated, technology changes rapidly, and they may find that they cannot stand still. Often, performance deficiencies occur or grow while organizational leadership is busily engaged in daily operations.

Leaders, like those identified in the three organizations above, can and have grown highly successful organizations. They made a very basic decision early. They realized that the strategic direction of their organization and the supporting performance improvements were *their responsibility.*

PART IV

INDEX

INDEX

A

B

E

F

H

J

K

L

N

O

P

Q

R

S

T

U

V

W